# 10 Virginia SOL Grade 7 Math Practice Tests

*The Ultimate Test Prep Collection with Answer Explanations*

**Dr. A. Nazari**

Copyright © 2026 Dr. A. Nazari

Published by View Math Education

ViewMath.com

All rights reserved. No part of this publication may be reproduced, distributed, or transmitted in any form or by any means, including photocopying, recording, or other electronic or mechanical methods, without the prior written permission of the author, except in the case of brief quotations embodied in critical reviews and certain other noncommercial uses permitted by copyright law, including Section 107 or 108 of the 1976 United States Copyright Act.

The information in this book is distributed on an "as is" basis, without warranty. While every precaution has been taken in the preparation of this work, neither the author nor the publisher shall have any liability to any person or entity with respect to any loss or damage caused or alleged to be caused directly or indirectly by the information contained in this book.

*Copyright © 2026*

# 10 Practice Tests

Grade 7 Math — Engineered for Mastery

Welcome to **The Architect's Workshop.**

Ten practice tests. Ten floors of a tower you're about to build. Each test lays another level of understanding — from foundation to capstone.

- **Foundation (Tests 1–3):** Learn the blueprints
- **Framework (Tests 4–6):** Build structural strength
- **Finishing (Tests 7–9):** Refine under pressure
- **Capstone (Test 10):** The final inspection

Precision. Practice. Perfection.

> **“** Every great structure
> is built one floor at a time.
> Ten levels of practice makes
> your math rock-solid. **”**

#  The Architectural Plan

A 4-phase construction plan for complete mastery

## Phase I: Foundation (Tests 1–3)

Untimed. Explore the test format and question types. Read every answer explanation after each test. Goal: understand the blueprint before you build.

Add a timer (70 minutes). Focus on the topics that gave you trouble in Phase I. Practice showing complete work and labeling units. Goal: build structural strength.

Full timed conditions (60 minutes). Simulate the real exam. Review only the questions you missed — don't re-study what you already know. Goal: refine under real pressure.

Your final inspection. Full exam conditions. This is the capstone — compare with Test 1 and measure your entire growth arc. Goal: prove your mastery.

**Multiple Choice** — select the single best answer

**Multi-Select** — choose ALL that apply

**Short Answer** — show your process

**Open Response** — explain and justify

# 📏 Engineering Principles 📏



## 📐 The D.R.A.F.T. Method

**D**   **Define**    What exactly is the question asking? Write it in your own words.

**R**   **Retrieve**    Pull out given information. List numbers, units, and relationships.

**A**   **Assemble**    Choose the right formula or strategy. Set up the equation or proportion.

**F**   **Figure**    Compute step by step. Show every operation. Label units.

**T**   **Test**    Does the answer make sense? Re-read the question. Verify the units.

## 📏 Seven Precision Rules

1. **Read twice, solve once.** The first read gives context; the second reveals what to compute.

3. **Estimate before calculating.** A quick mental approximation catches major errors.

5. **Track your signs.** Rational number operations are the #1 error source in Grade 7.

7. **Never submit blanks.** On open response, even a partial setup can earn credit.

## ⏱ Timing Blueprint

Tests 1–3: **Untimed** (learn the blueprints)     Tests 4–6: **70 min** (build speed)     Tests 7–10: **60 min** (exam conditions)

*Every great structure starts with a solid plan. Study the blueprints, learn from each test, and build something extraordinary.*

Get Online

Find more at
ViewMath.com/VA-Grade7

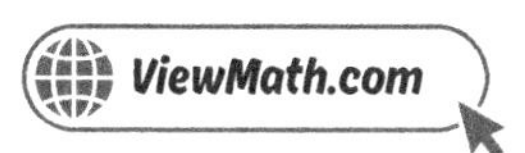

# The Drafting Toolkit

Prepare your station before every construction session



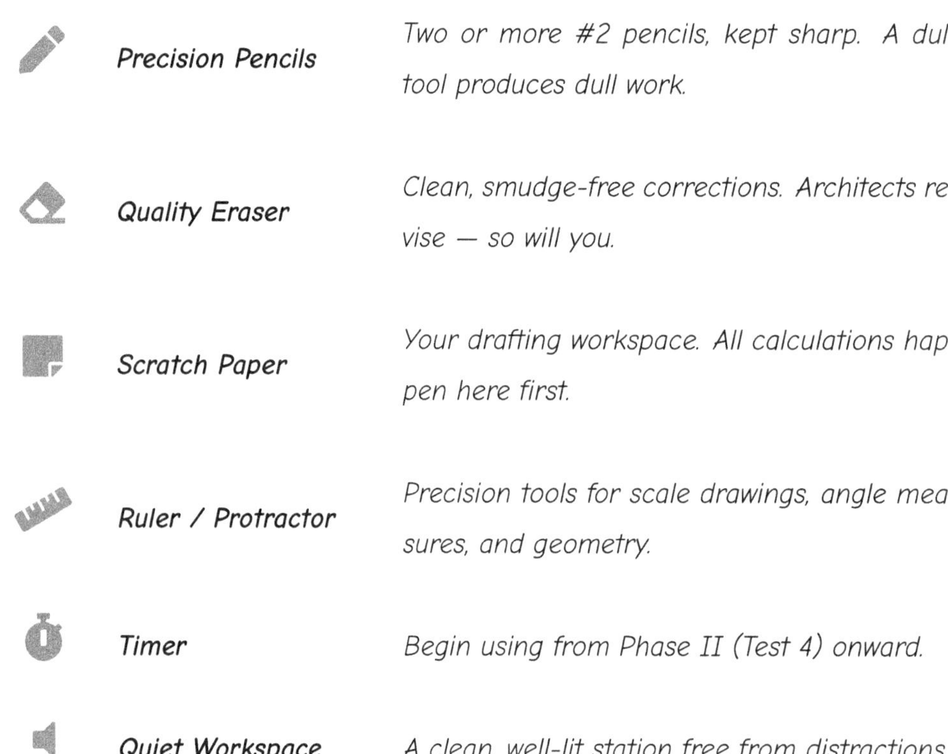

| | | |
|---|---|---|
| | **Precision Pencils** | Two or more #2 pencils, kept sharp. A dull tool produces dull work. |
| | **Quality Eraser** | Clean, smudge-free corrections. Architects revise — so will you. |
| | **Scratch Paper** | Your drafting workspace. All calculations happen here first. |
| | **Ruler / Protractor** | Precision tools for scale drawings, angle measures, and geometry. |
| | **Timer** | Begin using from Phase II (Test 4) onward. |
| | **Quiet Workspace** | A clean, well-lit station free from distractions. |

✓ Pencil and eraser

✓ Scratch paper (provided on test day)

✓ Ruler or protractor (if specified)

✓ Reference sheet (if provided)

✗ Calculator (unless specified)

✗ Electronic devices

✗ Notes or reference materials

✗ Communication with others

Ten practice tests provide the most comprehensive preparation available. The 4-phase structure (Foundation → Framework → Finishing → Capstone) ensures skills build progressively and durably.

*Keys to a successful build:*

- Space tests 2–3 days apart — never more than one per day
- Phases I–II: review answers together and discuss strategies
- Phases III–IV: let them work independently, then debrief results
- Compare Test 1 with Test 10 to celebrate the full construction arc

Find more at
ViewMath.com/VA-Grade7

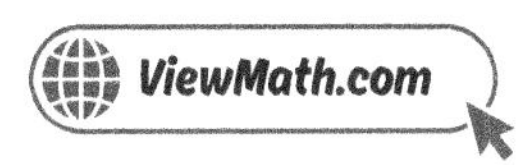

| Symbol | Name | What It Means | |
|--------|------|---------------|---|
| $(\ )$ | Parentheses | Do this part first. | $(3+4) \times 2 = 14$ |
| $10^3$ | Exponent | Multiply the base by itself that many times. $10^3 = 1{,}000$ | |
| $\frac{a}{b}$ | Fraction | $a$ parts out of $b$ equal parts; also means $a \div b$. | |
| $\frac{7}{3}$ | Improper Fraction | Numerator $\geq$ denominator. | $\frac{7}{3} = 2\frac{1}{3}$ |
| $0.45$ | Decimal | A number with a decimal point. | $0.45 = \frac{45}{100}$ |
| $>\ <\ =$ | Comparison | Greater than, less than, equal to. | $0.5 > 0.35$ |
| $(3,5)$ | Ordered Pair | A point on the coordinate plane: $(x, y)$. | |

## Key Formulas

- **Volume of a rectangular prism:**

  $V = l \times w \times h$

- **Order of operations:**

  Parentheses $\rightarrow$ Exponents $\rightarrow$ Multiply/Divide $\rightarrow$ Add/Subtract

- **Powers of 10:**

  $10^1 = 10 \quad 10^2 = 100$

  $10^3 = 1{,}000 \quad 10^4 = 10{,}000$

- **Fraction as division:**

  $\frac{a}{b} = a \div b$

## Place Value Chart

| Millions | 1,000,000 |
| --- | --- |
| Hundred-Thousands | 100,000 |
| Ten-Thousands | 10,000 |
| Thousands | 1,000 |
| Hundreds | 100 |
| Tens | 10 |
| Ones | 1 |

**Decimals**

| Tenths | 0.1 |
| --- | --- |
| Hundredths | 0.01 |
| Thousandths | 0.001 |

*Each place is 10× the place to its right, and $\frac{1}{10}$ of the place to its left.*

- **Sum** — the result of addition
- **Difference** — the result of subtraction
- **Product** — the result of multiplication
- **Quotient** — the result of division
- **Remainder** — what's left over after dividing
- **Factor** — a number you multiply
- **Expression** — numbers and operations without =
- **Equation** — a math sentence with =

- **Numerator** — the top number of a fraction
- **Denominator** — the bottom number of a fraction
- **Mixed number** — a whole number + a fraction
- **Equivalent fractions** — fractions with equal value
- **Decimal** — a number written with a decimal point
- **Volume** — the space inside a 3D shape
- **Coordinate plane** — a grid with $x$ and $y$ axes
- **Ordered pair** — $(x, y)$ location on the plane

- **Add** (+): total, altogether, combined, sum, increase, more than
- **Subtract** (−): difference, how many more, fewer, remain, decrease, left

- **Multiply** (×): each, every, groups of, times, product, per, of (with fractions)
- **Divide** (÷): share equally, split, each group, how many groups, quotient, per

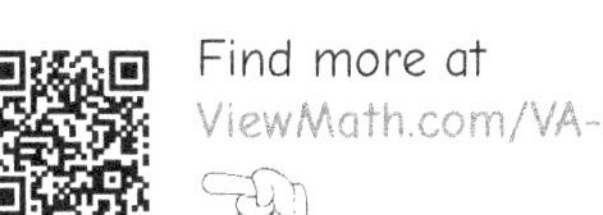

Find more at
ViewMath.com/VA-Grade7

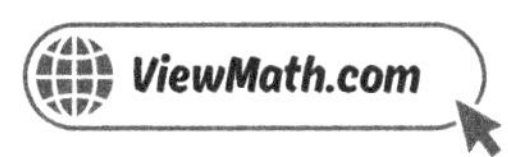

# Multiplication Table

| × | 1 | 2 | 3 | 4 | 5 | 6 | 7 | 8 | 9 | 10 | 11 |
|---|---|---|---|---|---|---|---|---|---|----|----|
| 1 | 1 | 2 | 3 | 4 | 5 | 6 | 7 | 8 | 9 | 10 | 11 |
| 2 | 2 | 4 | 6 | 8 | 10 | 12 | 14 | 16 | 18 | 20 | 22 |
| 3 | 3 | 6 | 9 | 12 | 15 | 18 | 21 | 24 | 27 | 30 | 33 |
| 4 | 4 | 8 | 12 | 16 | 20 | 24 | 28 | 32 | 36 | 40 | 44 |
| 5 | 5 | 10 | 15 | 20 | 25 | 30 | 35 | 40 | 45 | 50 | 55 |
| 6 | 6 | 12 | 18 | 24 | 30 | 36 | 42 | 48 | 54 | 60 | 66 |
| 7 | 7 | 14 | 21 | 28 | 35 | 42 | 49 | 56 | 63 | 70 | 77 |
| 8 | 8 | 16 | 24 | 32 | 40 | 48 | 56 | 64 | 72 | 80 | 88 |
| 9 | 9 | 18 | 27 | 36 | 45 | 54 | 63 | 72 | 81 | 90 | 99 |
| 10 | 10 | 20 | 30 | 40 | 50 | 60 | 70 | 80 | 90 | 100 | 110 |
| 11 | 11 | 22 | 33 | 44 | 55 | 66 | 77 | 88 | 99 | 110 | 121 |

## How to Use This Table

To find **4 × 7**:

1. Find **4** in the left column (blue).
2. Find **7** in the top row (blue).
3. Follow the row and column until they meet: the answer is **28**!

**Tip:** You can also use this table for division! If you know $28 \div 4 = $ ?, find 28 in the 4's row. The column header gives you the answer: **7**!

# 🏢 Construction Log 🏢

Architect: _________________________________          Project Start: ___________________

🅰 Floor 1          Date: ___________          Score: _____ / _____          %: _____

🅰 Floor 2          Date: ___________          Score: _____ / _____          %: _____

🅰 Floor 3          Date: ___________          Score: _____ / _____          %: _____


🅰 Floor 4          Date: ___________          Score: _____ / _____          %: _____

🅰 Floor 5          Date: ___________          Score: _____ / _____          %: _____

🅰 Floor 6          Date: ___________          Score: _____ / _____          %: _____


🅰 Floor 7          Date: ___________          Score: _____ / _____          %: _____

🅰 Floor 8          Date: ___________          Score: _____ / _____          %: _____

🅰 Floor 9          Date: ___________          Score: _____ / _____          %: _____


🅰 Floor 10          Date: ___________          Score: _____ / _____          %: _____

**Certification Level:**  Apprentice (0–39%)   Technician (40–59%)   Engineer (60–79%)   **Master Architect** (80–100%)

YOUR ONLINE COMPANION

# Continue Learning at
# ViewMath Academy!

## For Parents, Teachers & Students

Great job on the practice tests! Want to keep improving? ViewMath Academy is your free online companion to this book.

- **Score Analyzer** — Enter your answers and instantly see which topics need more practice

- **Interactive Lessons** — Review the concepts behind each question with clear explanations

- **Adaptive Quizzes** — Practice your weak topics with questions that match your level

- **Progress Tracking** — See your mastery grow across all Grade 7 math topics

- **Personalized Dashboard** — A learning plan tailored just for you

Scan to visit ViewMath Academy

ViewMath.com/VA-Grade7

# Table of Contents

*Here's what we'll explore together!*

*Let's learn and have fun!*

# 1

# Practice Test 1

✅ 30 Questions

## ✏️ Before You Start ✏️

- ✔️ **Read each question carefully** before choosing your answer.
- ✔️ **Show your work** on scratch paper when you need to.
- ✔️ **Skip hard questions** and come back to them later.
- ✔️ **Check your answers** when you're done.
- ✔️ **Take your time** — there's no rush!

⭐ You've Got This! ⭐

A store sells beads at a constant price. The equation is $c = 0.15b$. How many beads can you buy with \$6?

The bar graph shows how many students chose each favorite sport. What percent of the students chose soccer?

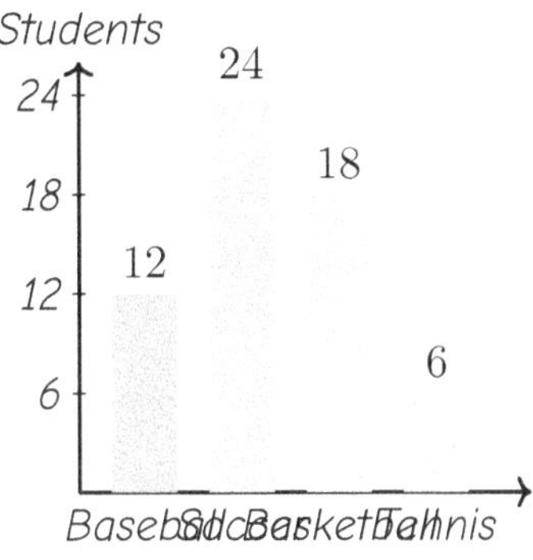


A   24%                                                    B   30%

C   40%                                                    D   50%

3. Look at the proportion model below. What value of $n$ makes the proportion true?

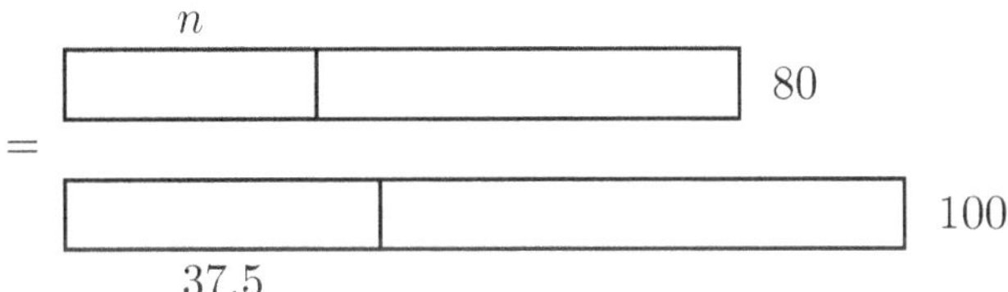


A   28                                                     B   30

C   32                                                     D   37.5

4. A plant was 8 inches tall. It grew to 14 inches. What is the percent increase?

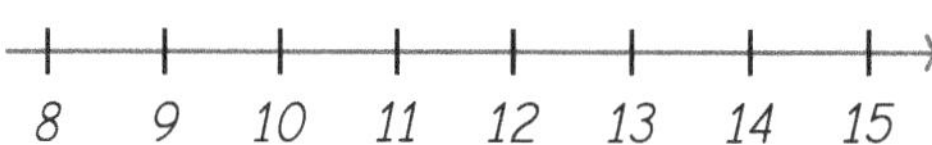

A) 43%                                   B) 57%

C) 60%                                   D) 75%

5. The number line below shows an estimated value and an actual value. What is the percent error?

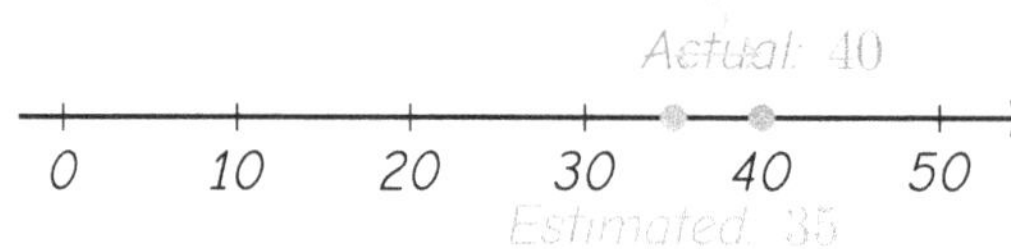

A) 5%                                    B) 10%

C) 12.5%                                 D) 14.3%

6. What integer must you add to $-15$ to get $-6$?

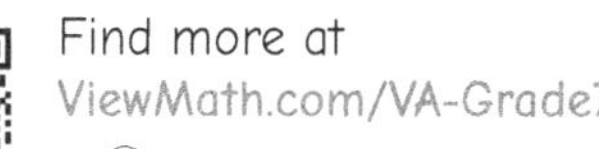
Get Online

Find more at
ViewMath.com/VA-Grade7

*The diagram shows elevations. What is the difference in elevation from Point B to Point A?*

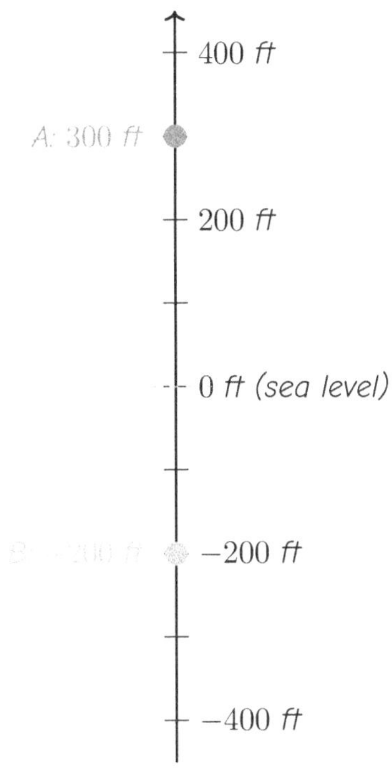


A   100 ft

B   −100 ft

C   500 ft

D   −500 ft

*What is $5.6 - 8.1$?*

A   2.5

B   −2.5

C   13.7

D   −13.7

9. *Expand and simplify $-2(x + 6) + 3(2x - 1)$.*

Get Online

Find more at
ViewMath.com/VA-Grade7

ViewMath.com

10. *Factor* $-10m - 15.$

11. *The thermometer shows the current temperature. The temperature was* $-3.5°F$ *this morning and rose t degrees each hour for 4 hours to reach the current reading. What is t?*

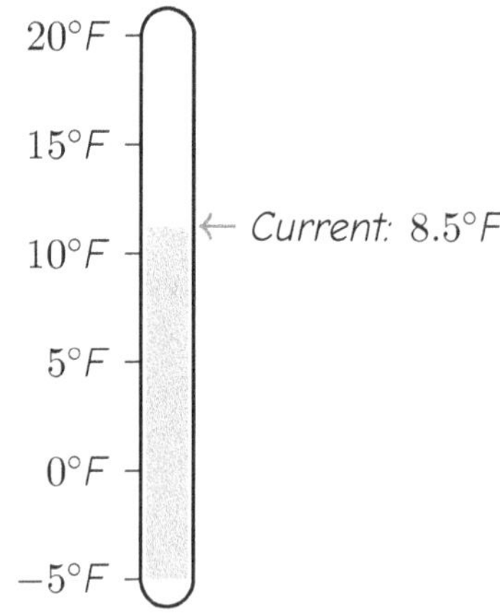


A) $t = 3$          B) $t = 5$

C) $t = 1.25$        D) $t = 2$

12. *The table shows ticket prices for groups. A school has $200 to spend and must also pay a $35 bus fee. Write and solve an inequality to find the maximum number of students s who can go.*

| Group Size | Price per Student |
|------------|-------------------|
| 1–10 | $12 |
| 11–25 | $9 |
| 26+ | $7 |

*Assume the group qualifies for the $9 rate.*

13. A phone battery is at 85% and loses 7% per hour. Write an inequality for when the battery is below 20%, and solve.

14. A map uses the scale $1\ cm = 25\ km$. Two towns are 6 cm apart on the map. What is the actual distance?

   A) 31 km                                B) 125 km

   C) 150 km                              D) 175 km

15. A drawing uses $1\ cm = 7\ m$. You redraw at $1\ cm = 7\ m$ (the same scale). A line is 5 cm long. How long is it on the new drawing?

16. You slice a sphere with any flat plane. What shape is the cross-section?

   A) Oval                                   B) Rectangle

   C) Circle                                D) It depends on the angle of the cut

17. Two supplementary angles are in the ratio $5 : 4$. Find both angles.

18. Rotate point $Q(4, -3)$ by $90°$ clockwise about the origin. What are the coordinates of $Q'$?

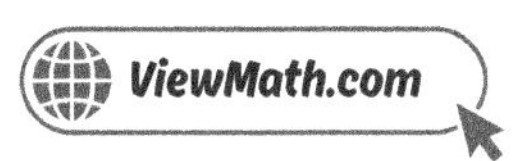

19. Are all circles similar to each other?

- A) No, they have different radii
- B) No, only congruent circles are similar
- C) Yes, all circles are similar
- D) Only if they have the same center

20. A pentagon can be divided into a rectangle 6 cm by 4 cm and a triangle with base 6 cm and height 2 cm. What is the total area?

- A) $24\ cm^2$
- B) $30\ cm^2$
- C) $36\ cm^2$
- D) $48\ cm^2$

21. A triangular prism has a triangular base with base 6 cm and height 4 cm, and the prism is 10 cm long. The three rectangular faces have widths 6 cm, 5 cm, and 5 cm. What is the total surface area?

- A) $184\ cm^2$
- B) $160\ cm^2$
- C) $124\ cm^2$
- D) $120\ cm^2$

22. A cube has a volume of $64\ cm^3$. What is the side length?

- A) 2 cm
- B) 4 cm
- C) 8 cm
- D) 16 cm

23. What is the surface area of a cylinder with radius 5 cm and height 8 cm? Use $\pi \approx 3.14$

- A) $251.2\ cm^2$
- B) $314\ cm^2$
- C) $408.2\ cm^2$
- D) $628\ cm^2$

24. In a school of 600 students, a random sample of 60 found that 15 students walk to school. Predict how many students in the entire school walk to school.

A  100

B  150

C  200

D  250

25. Which measure of variability is shown directly on a box plot?

A  Mean absolute deviation (MAD)

B  Standard deviation

C  Interquartile range (IQR)

D  Mean

26. The table compares two delivery companies' shipping times (in days).

|        | Company X | Company Y |
|--------|-----------|-----------|
| Mean   | 3.5 days  | 3.2 days  |
| Median | 3 days    | 3 days    |
| Range  | 4 days    | 8 days    |
| IQR    | 1.5 days  | 4 days    |

A customer wants the most reliable delivery time. Which company should they choose?

A  Company Y, because its mean is lower

B  Company X, because its IQR and range are both smaller

C  Company Y, because its range includes more possibilities

D  They are equally reliable

27. In a stem-and-leaf plot, 8 | 0 represents:

A  8

B  0.8

C  80

D  8.0

Get Online

Find more at
ViewMath.com/VA-Grade7

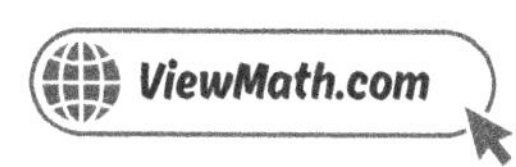
ViewMath.com

28. A die is rolled 180 times. The number 4 appears 25 times. What is the experimental probability of rolling a 4? Write your answer as a fraction in simplest form.

29. A spinner has 4 equal sections $(1, 2, 3, 4)$ and a die is rolled. What is the probability that both show a 3?

A) $\frac{1}{10}$

B) $\frac{1}{24}$

C) $\frac{2}{10}$

D) $\frac{1}{4}$

30. A spinner has a 60% chance of landing on blue. In a simulation of 80 spins, how many times would you expect blue to appear?

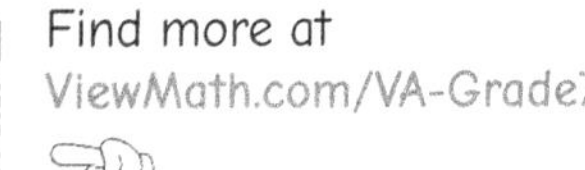

# ⭐ *End of Practice Test 1* ⭐

*Great job finishing the test!*

 *My Score*

*I got _____________ out of 30 questions right.*

*📊 Check Your Score Online!*

*Visit **ViewMath Academy** to enter your answers and see which topics you need to review. You can also explore lessons, take quizzes, track your scores, and save your progress!*

*viewmath.com/score/7.1.VA.16*

2

# Practice Test 2

📋 30 Questions

✏️ Before You Start ✏️

- ✓ **Read each question carefully** before choosing your answer.
- ✓ **Show your work** on scratch paper when you need to.
- ✓ **Skip hard questions** and come back to them later.
- ✓ **Check your answers** when you're done.
- ✓ **Take your time** — there's no rush!

⭐ You've Got This! ⭐

Do your best and show what you know!

The table below shows a proportional relationship. Using the equation, what is $y$ when $x = 9$?

| $x$ | $y$ |
|---|---|
| 2 | 11 |
| 5 | 27.5 |
| 7 | 38.5 |

A  $y = 45$

B  $y = 49.5$

C  $y = 54$

D  $y = 63$

Look at the $10 \times 10$ grid below. What percent of the grid is **not** shaded?

3. The model below shows that a number $w$ is split into a shaded part and an unshaded part. The shaded part is 84 and represents 70% of the whole. What is the whole number $w$?

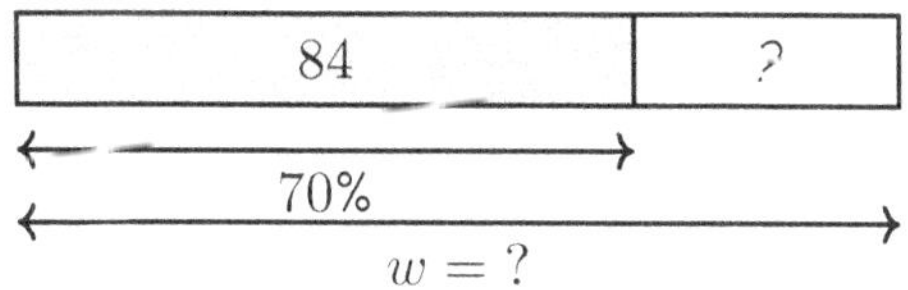

Find more at
ViewMath.com/VA-Grade7

ViewMath.com

4. What multiplier represents a 25% increase?

A) 0.25

B) 0.75

C) 1.25

D) 25

5. An estimate has a percent error of 0%. What does this mean?

A) The estimate was very close

B) The estimate was exactly correct

C) The actual value was 0

D) There was no actual value to compare

6. What is $(-7) + 7 + (-3)$?

A) $-3$

B) $3$

C) $-17$

D) $17$

7. What is $(-7) - (-12)$?

A) $-19$

B) $19$

C) $-5$

D) $5$

8. What is $-\dfrac{5}{6} - \left(-\dfrac{1}{2}\right)$?

Get Online

Find more at
ViewMath.com/VA-Grade7

9. Expand $\frac{1}{2}(8x - 6)$.

A  $4x - 6$          B  $8x - 3$

C  $4x - 3$          D  $4x + 3$

10. Factor $36a + 24b$.

11. Solve $\dfrac{5x}{2} + 3 = 18$.

A  $x = 3$          B  $x = 6$

C  $x = 42$          D  $x = 15$

12. Solve $8 - 3x \geq 23$.

13. A student has test scores of 78, 85, and 90. She needs an average of at least 85 on four tests to earn an A. Solve the inequality to find the minimum score $s$ on the fourth test, and describe the graph.

$$\xleftarrow{\phantom{x}}\!\!\!+\!\!+\!\!+\!\!+\!\!+\!\!+\!\!+\!\!+\!\!+\!\!\xrightarrow{\phantom{x}}$$
65 70 75 80 85 90 95 100

Get Online

Find more at
ViewMath.com/VA-Grade7

ViewMath.com

14. *A drawing of a room uses scale $1\ cm = 3\ m$. The drawing shows a square room with side $4\ cm$. If the scale factor doubles all lengths, by what factor does the actual area change?*

15. *When you redraw a scale drawing at a smaller scale (more real units per cm), the new drawing is:*

   A. *Larger than the original drawing*    B. *Smaller than the original drawing*

   C. *The same size as the original drawing*    D. *A different shape than the original drawing*

16. *You slice a cube with a vertical cut from one face to the opposite face, perpendicular to the base. What is the most likely cross-section shape?*

   A. *Square or rectangle*    B. *Circle*

   C. *Triangle*    D. *Hexagon*

17. *Which pair of angles are supplementary?*

   A. $45°$ *and* $45°$    B. $60°$ *and* $30°$

   C. $110°$ *and* $70°$    D. $100°$ *and* $100°$

18. *Point $H(-3, 7)$ is reflected over the $y$-axis. What are the coordinates of $H'$?*

   A. $(-3, -7)$    B. $(3, 7)$

   C. $(3, -7)$    D. $(7, -3)$

Find more at
ViewMath.com/VA-Grade7

A tree casts a 16-ft shadow. A 4-ft fence post casts a 5-ft shadow at the same time. How tall is the tree? Round to the nearest tenth.


A shape is made of a square with side 8 cm and a semicircle with diameter 8 cm attached to one side. What is the area? Use $\pi \approx 3.14$.

A  64 $cm^2$                                          B  89.12 $cm^2$

C  114.24 $cm^2$                                      D  139.12 $cm^2$


21. A paint company needs to paint the outside of a box that is 3 m by 2 m by 1 m. Each can of paint covers 5 $m^2$. How many cans are needed?

A  3                                                  B  4

C  5                                                  D  6


A planter box is 3 ft long, 1.5 ft wide, and 2 ft deep. How many cubic feet of soil are needed to fill it?


Which part of the surface area formula $SA = 2\pi r^2 + 2\pi rh$ represents the area of the two circular bases?

A  $2\pi rh$                                          B  $2\pi r^2$

C  $\pi r^2 h$                                        D  $\pi r^2$

Find more at
ViewMath.com/VA-Grade7

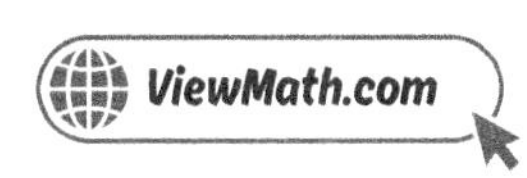

24. *A factory produces 10,000 batteries per day. A sample of 200 batteries is tested, and 8 are defective. What percent of the sample is defective?*

25. *The dot plots below show the number of push-ups completed by students in two gym classes.*

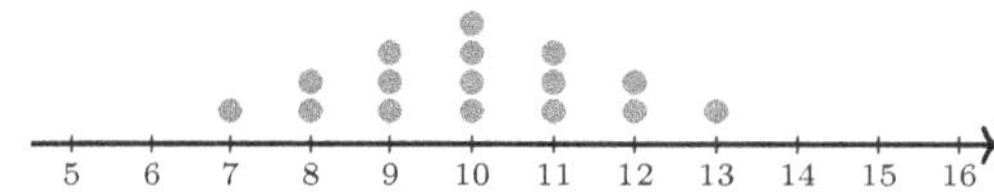

*Which class has a higher center, and by approximately how many push-ups?*

A  *Class 1, by about 3*                    B  *Class 2, by about 3*

C  *Class 2, by about 6*                    D  *They have the same center*

26. *The dot plots show daily high temperatures (°F) for two cities over 10 days.*

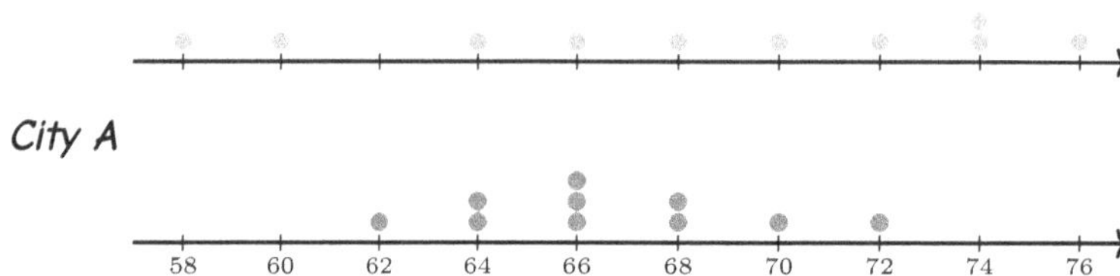

*Find the mean and range for each city. Which city has more consistent temperatures?*

Get Online

Find more at
ViewMath.com/VA-Grade7

27. In a stem-and-leaf plot, 5 | 3 means:

A  5.3                                              B  35

C  53                                              D  $5 + 3 = 8$


28. A die is rolled 120 times. Each number appears the following number of times: 1: 22, 2: 18, 3: 20, 4: 21, 5: 19, 6: 20. Which number had the highest experimental probability?

A  1                                               B  3

C  4                                               D  6


29. The tree diagram shows the outcomes for flipping two coins. What is the probability of getting at least one tail?

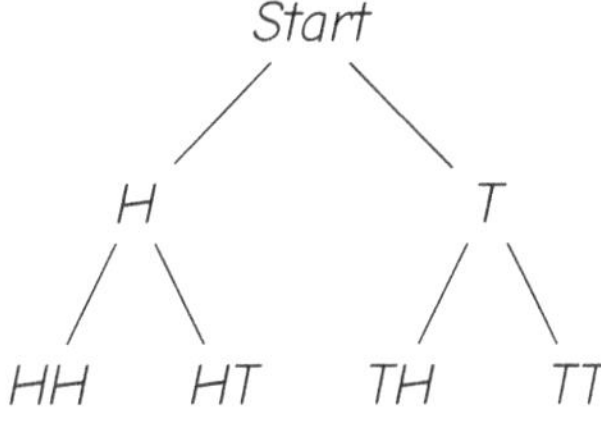


A  $\frac{1}{4}$                                              B  $\frac{1}{2}$

C  $\frac{3}{4}$                                              D  $\frac{2}{4}$


Get Online
Find more at
ViewMath.com/VA-Grade7

30. *The bar graph shows the results of simulating a spinner 100 times. The spinner is supposed to land on each color with equal probability. Based on the simulation, does the spinner appear fair?*

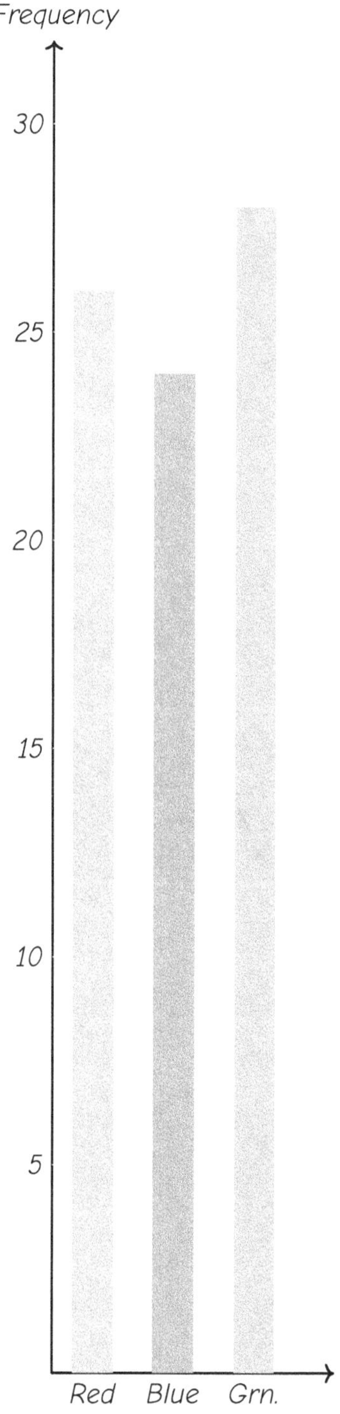


*Not shown: Yellow = 22*

(A) *No, because the bars are different heights*

(B) *Yes, the frequencies $(26, 24, 28, 22)$ are all roughly close to $25$*

(C) *Green is clearly the most likely color*

(D) *Cannot be determined from the graph*

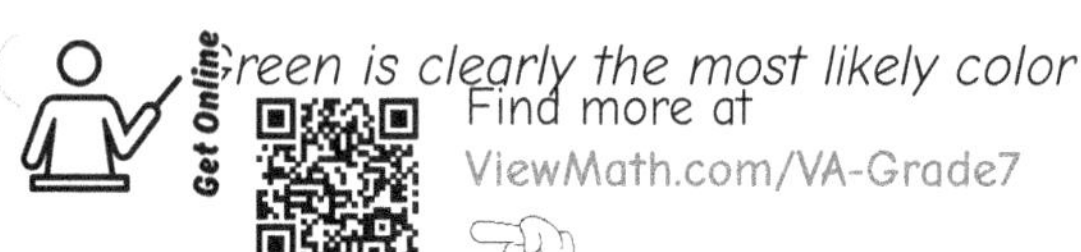

 # End of Practice Test 2 

*Great job finishing the test!*

 My Score

*I got _____________ out of 30 questions right.*

Check Your Score Online!

*Visit **ViewMath Academy** to enter your answers and see which topics you need to review. You can also explore lessons, take quizzes, track your scores, and save your progress!*

viewmath.com/score/7.1.VA.17

# 3

# Practice Test 3

☑ *30 Questions*

## ✏ Before You Start ✏

- ✓ **Read each question carefully** before choosing your answer.
- ✓ **Show your work** on scratch paper when you need to.
- ✓ **Skip hard questions** and come back to them later.
- ✓ **Check your answers** when you're done.
- ✓ **Take your time** — there's no rush!

⭐ You've Got This! ⭐

*The table shows a proportional relationship. Write the equation and find $y$ when $x = 15$.*

| $x$ | $y$ |
|---|---|
| 2 | 7 |
| 4 | 14 |
| 6 | 21 |

A  $y = 3.5x;\ y = 52.5$         B  $y = 7x;\ y = 105$

C  $y = 2x;\ y = 30$         D  $y = 3.5x;\ y = 45$

*A class has 32 students. If 75% passed the test, how many students passed?*

A  20         B  24

C  25         D  28

3. *Sarah and Jake both solve "What is 40% of 75?" Sarah uses a proportion; Jake uses an equation. Which statement is true?*

A  Only the proportion method works         B  Only the equation method works

C  Both methods give the same answer         D  The methods give different answers

*A town's population was 12,000 last year and is 13,200 this year. What is the percent increase?*

A  8%         B  10%

C  12%         D  15%

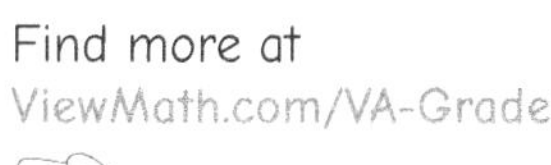


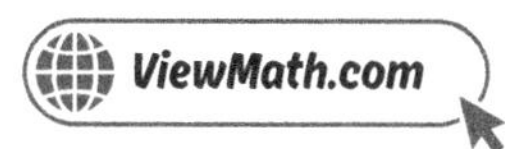

5. A carpenter estimated a board was 8 feet long.  The actual length was 8.5 feet.  What is the percent error (rounded to the nearest tenth)?

A) 5%

B) 5.9%

C) 6.3%

D) 6.7%

6. What is $(-12) + 7$?

A) $-19$

B) 19

C) $-5$

D) 5

7. Death Valley's lowest point is $-282$ feet.  A nearby hill is 134 feet above sea level.  What is the difference in elevation?

A) 148 feet

B) $-148$ feet

C) 416 feet

D) $-416$ feet

8. What is $4.2 + (-6.7)$?

A) 10.9

B) $-10.9$

C) 2.5

D) $-2.5$

9. Which expression is equivalent to $7(a - 3) - 2a$?

A) $5a - 3$

B) $5a + 21$

C) $9a - 21$

D) $5a - 21$

Get Online

Find more at
ViewMath.com/VA-Grade7

ViewMath.com

Factor $20n - 30$.

A. $5(4n - 6)$

B. $10(2n - 3)$

C. $2(10n - 15)$

D. $20(n - 30)$

Solve $-1.5x + 4.5 = -3$.

A. $x = -5$

B. $x = 1$

C. $x = 5$

D. $x = -1$

12. Solve $-7x - 4 < 17$.

Solve $-2x \leq 10$ and describe the graph.

A. Closed circle at $-5$, shade left

B. Closed circle at $5$, shade left

C. Closed circle at $-5$, shade right

D. Open circle at $-5$, shade right

A blueprint uses $1\ in = 9\ ft$. A room is $45\ ft$ long in real life. How long is the room on the blueprint?

15. Original scale: $1\ cm = 10\ m$. New scale: $1\ cm = 40\ m$. What is the scale ratio?

A. $\frac{1}{4}$

B. $4$

C. $30$

D. $\frac{1}{30}$

Find more at
ViewMath.com/VA-Grade7

ViewMath.com

16. A triangular prism is sliced with a vertical cut perpendicular to the triangular bases. What shape is the cross-section?

   A  Triangle                     B  Rectangle

   C  Circle                       D  Trapezoid

17. In the diagram below, lines $AB$ and $CD$ intersect at point $O$. Find the value of $y$.

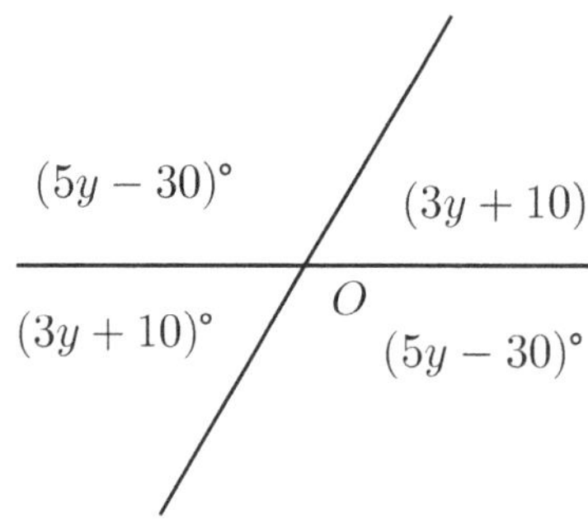


   A  10                     B  20

   C  25                     D  35

18. A triangle has vertices $P(1,2)$, $Q(4,2)$, and $R(4,6)$. After a translation, $P' = (3,0)$. What translation was applied?

   A  Right 2, down 2               B  Left 2, up 2

   C  Right 2, up 2                 D  Left 2, down 2

19. A 5-ft child stands next to a wall that casts a 15-ft shadow. The child casts a 3-ft shadow. How tall is the wall?

   A  9 ft                     B  25 ft

   C  45 ft                   D  10 ft

Find more at
ViewMath.com/VA-Grade7

20. A T-shaped figure can be broken into a 12 cm by 2 cm horizontal rectangle and a 2 cm by 8 cm vertical rectangle. What is the total area?

A  24 cm$^2$                                      B  32 cm$^2$

C  40 cm$^2$                                      D  48 cm$^2$

21. If all dimensions of a rectangular prism are doubled, what happens to its surface area?

A  It doubles                                     B  It triples

C  It quadruples                                  D  It increases by a factor of 8

22. A composite solid is made of two rectangular prisms joined together as shown. What is the total volume?

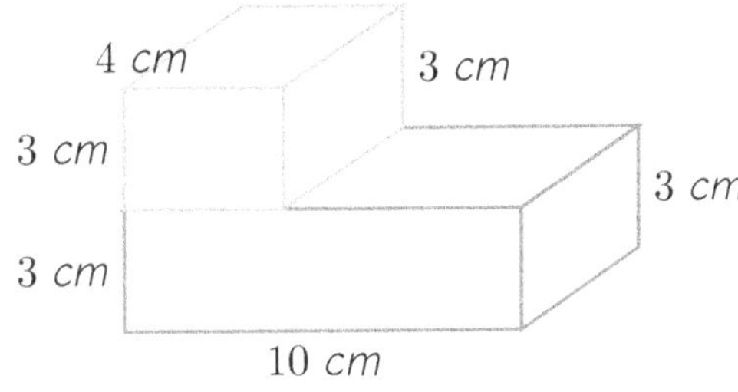


23. If you double the height of a cylinder but keep the radius the same, what happens to the volume?

A  It stays the same                              B  It doubles

C  It quadruples                                  D  It increases by a factor of 8

24. Which of the following increases the reliability of conclusions drawn from a sample?

A  Using a smaller sample                         B  Using a larger random sample

C  Surveying only people who agree with you       D  Using a non-random sample

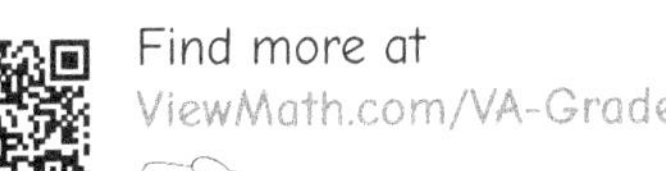
Find more at
ViewMath.com/VA-Grade7

25   *School A: median score 88, MAD 4. School B: median score 82, MAD 4. The difference in medians expressed in MADs is:*

  (A)  1 MAD

  (B)  1.5 MADs

  (C)  2 MADs

  (D)  4 MADs

26   *Group X: mean 45, MAD 6. Group Y: mean 57, MAD 6. Express the difference in means as a multiple of MAD.*

27.  *Why is it important to include a key (e.g., "3 | 4 means 34") in a stem-and-leaf plot?*

  (A)  It makes the plot look nicer

  (B)  It tells the reader how to interpret the stems and leaves

  (C)  It is not important

  (D)  It replaces the title

28.  *A spinner lands on blue 18 times out of 90 spins. What is the experimental probability of landing on blue?*

  (A)  $\frac{1}{5}$

  (B)  $\frac{1}{4}$

  (C)  $\frac{18}{90}$

  (D)  Both A and C

29.  *Two dice are rolled. What is the probability of getting a sum of 2?*

  (A)  $\frac{1}{6}$

  (B)  $\frac{1}{12}$

  (C)  $\frac{1}{36}$

  (D)  $\frac{2}{36}$

Find more at
ViewMath.com/VA-Grade7

30. You simulate flipping a coin 3 times and counting how many times all 3 are heads. In 40 simulations, all 3 heads occurs 6 times. What is the experimental probability?

A  $\frac{6}{40} = 0.15$

B  $\frac{1}{8} = 0.125$

C  $\frac{3}{40} = 0.075$

D  $\frac{6}{120} = 0.05$

Find more at
ViewMath.com/VA-Grade7

Get Online

ViewMath.com

#  End of Practice Test 3 

*Great job finishing the test!*

 **My Score**

I got _____________ out of 30 questions right.

Check your answers in the Answer Key at the back of the book.

Review any questions you missed. That's how we learn!

##  Check Your Score Online!

*Visit **ViewMath Academy** to enter your answers and see which topics you need to review. You can also explore lessons, take quizzes, track your scores, and save your progress!*

*viewmath.com/score/7.1.VA.18*

Or go to viewmath.com/score and enter code: 7.1.VA.18

# 4

# Practice Test 4

☑ *30 Questions*

---

## ✏️ Before You Start ✏️

- ✓ **Read each question carefully** before choosing your answer.
- ✓ **Show your work** on scratch paper when you need to.
- ✓ **Skip hard questions** and come back to them later.
- ✓ **Check your answers** when you're done.
- ✓ **Take your time** — there's no rush!

★ You've Got This! ★

Do your best and show what you know!

A bike rental costs \$8 per hour. Which equation represents the total cost $c$ for $h$ hours?

A. $c = h + 8$

B. $c = 8h$

C. $c = \frac{h}{8}$

D. $c = 8 + h$

A parking lot has spaces for 350 cars. Currently 280 spaces are filled. What percent of the lot is full?

3. 135 is 54% of what number?

A. 72.9

B. 200

C. 245

D. 250

A pond had 400 fish last year and 340 this year. What is the percent decrease?

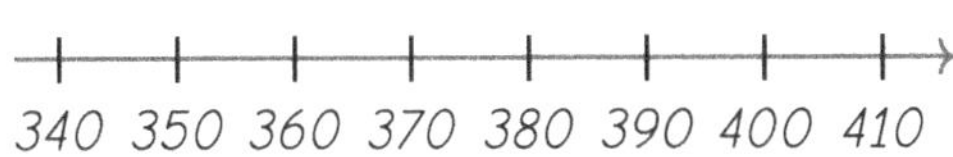


You estimated there were 50 marbles in a jar. The actual count was 40. What is the percent error?

A. 10%

B. 20%

C. 25%

D. 50%

Get Online

Find more at
ViewMath.com/VA-Grade7

ViewMath.com

6. What is $15 + (-15)$?

   (A) 30           (B) $-30$

   (C) 15           (D) 0

7. The high temperature was 4°F and the low was $-11$°F. What is the difference between the high and low?

   (A) 7°F           (B) $-7$°F

   (C) 15°F           (D) $-15$°F

8. What is $\dfrac{7}{10} - \dfrac{4}{5}$?

9. Expand and simplify $3(x - 2) + 2(x + 4)$.

   (A) $5x + 2$           (B) $5x - 2$

   (C) $5x + 14$           (D) $x + 2$

10. Factor $6x + 12$.

   (A) $2(3x + 6)$           (B) $6(x + 2)$

   (C) $3(2x + 4)$           (D) $6(x + 12)$

11. Solve $\dfrac{n}{2} + \dfrac{n}{3} = 10$.

   (A) $n = 15$           (B) $n = 12$

   (C) $n = 6$           (D) $n = 20$

Find more at
ViewMath.com/VA-Grade7

ViewMath.com

12. *Write an inequality: "A number $x$ divided by 4, minus 3, is at least 2." Then solve.*

*Which graph correctly shows the solution to $2x - 5 > 3$?*

**Graph A:**

**Graph B:**

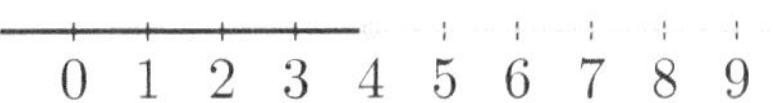

**Graph C:**

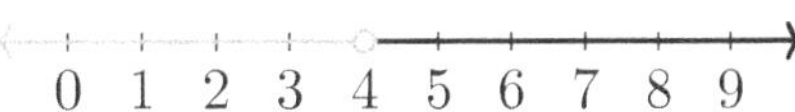

  A  *Graph A*        B  *Graph B*

  C  *Graph C*        D  *None of the above*

*The actual length of a bridge is $360$ m. On a scale drawing, the bridge is $12$ cm. What is the scale in the form "$1$ cm $=$ __ m"?*

15. *A map uses $1$ cm $= 20$ km. You redraw it at $1$ cm $= 10$ km. What happens to the drawing?*

  A  *Every length is halved*      B  *Every length stays the same*

  C  *Every length is doubled*      D  *Every length is multiplied by $10$*

Find more at
ViewMath.com/VA-Grade7

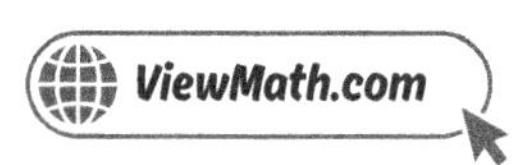

16. *The diagram shows a triangular prism. A cut is made parallel to the triangular base (shown as the shaded region). What is the cross-section?*

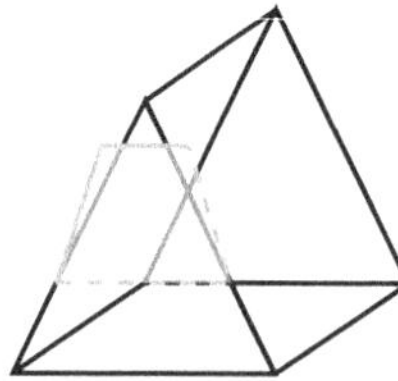

   A  Rectangle                       B  Triangle

   C  Circle                            D  Hexagon

17. *Angles $A$ and $B$ are complementary. Angle $A$ is $(3x-5)°$ and angle $B$ is $(x+15)°$. What is angle $A$?*

   A  20°                           B  35°

   C  55°                           D  65°

18. *After reflecting point $K$ over the $x$-axis, the image is $K'(4,-3)$. What were the original coordinates of $K$?*

   A  $(4,3)$                       B  $(-4,-3)$

   C  $(-4,3)$                   D  $(4,-3)$

19. *Triangle $LMN \sim$ Triangle $XYZ$. $LM=10$, $MN=14$, $LN=16$, $XY=5$. Find $YZ$.*

   A  7                           B  8

   C  9                           D  28

Find more at
ViewMath.com/VA-Grade7

ViewMath.com

A banner is shaped like a rectangle $18$ in by $6$ in with a triangle cut from one end. The triangle has a base of $6$ in and a height of $4$ in. What is the area of the banner?

A  $84\ in^2$

B  $96\ in^2$

C  $108\ in^2$

D  $120\ in^2$

21. How many faces does a rectangular prism have?

A  $4$

B  $5$

C  $6$

D  $8$

What unit is used to measure volume?

A  Square units $(cm^2)$

B  Cubic units $(cm^3)$

C  Linear units $(cm)$

D  Degrees

The diagram shows the net of a cylinder. What is the total surface area? Use $\pi \approx 3.14$.

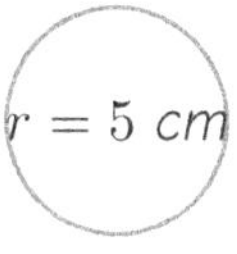

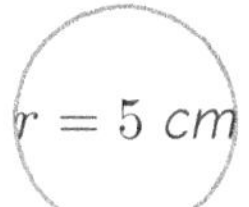


A  $157\ cm^2$

B  $376.8\ cm^2$

C  $533.8\ cm^2$

D  $691.6\ cm^2$


Get Online
Find more at
ViewMath.com/VA-Grade7

ViewMath.com

24. Give one reason why a sample might not perfectly represent a population.

25. When comparing two populations visually, you should look at:

A. Only the center (mean or median)

B. Only the spread (range or MAD)

C. Both the center and the spread

D. Only the largest and smallest values

26. Store A daily sales: mean $500, MAD $40. Store B daily sales: mean $620, MAD $40. The difference in means in MADs is:

A. 1.5 MADs

B. 3 MADs

C. 40 MADs

D. 120 MADs

27. A stem-and-leaf plot shows 12 data values. What is the median?

A. The 6th value

B. The 7th value

C. The average of the 6th and 7th values

D. The 12th value

28. As the number of trials in an experiment increases, what happens to the experimental probability?

A. It always equals the theoretical probability

B. It gets farther from the theoretical probability

C. It tends to get closer to the theoretical probability

D. It stays the same regardless of the number of trials

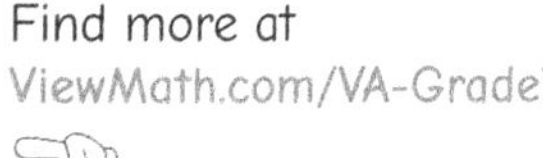

Find more at
ViewMath.com/VA-Grade7

29. Look at the tree diagram for flipping three coins. What is the probability of getting exactly 2 tails? Write your answer as a fraction in simplest form.

<pre>
                        Start
                     /         \
                 H                 T
               /   \             /   \
             H       T         H       T
            / \     / \       / \     / \
          HHH HHT HTH HTT   THH THT TTH TTT
</pre>

30. A student wants to simulate the probability that exactly 2 out of 3 students pass a test, where each student has a 50% chance. Which model would work?

A  Roll one die: odd = pass, even = fail

B  Flip 3 coins: heads = pass, tails = fail; count exactly 2 heads

C  Draw 3 marbles from a bag of 2 red and 1 blue

D  Flip 1 coin 3 times and count all tails

Get Online

Find more at
ViewMath.com/VA-Grade7

ViewMath.com

# ⭐ End of Practice Test 4 ⭐

*Great job finishing the test!*

 My Score

I got ____________ out of 30 questions right.

Check your answers in the Answer Key at the back of the book.

Review any questions you missed. That's how we learn!

## 📊 Check Your Score Online!

Visit **ViewMath Academy** to enter your answers and see which topics you need to review. You can also explore lessons, take quizzes, track your scores, and save your progress!

viewmath.com/score/7.1.VA.19

Or go to viewmath.com/score and enter code 7.1.VA.19

# 5

# Practice Test 5

✅ 30 Questions

## ✏️ Before You Start ✏️

- ✓ **Read each question carefully** before choosing your answer.
- ✓ **Show your work** on scratch paper when you need to.
- ✓ **Skip hard questions** and come back to them later.
- ✓ **Check your answers** when you're done.
- ✓ **Take your time** — there's no rush!

⭐ You've Got This! ⭐

1. *A graph passes through* $(5, 17.5)$. *Write the proportional equation and find* $x$ *when* $y = 49$.

2. 18 *is what percent of 72?*

A  18%                                          B  20%

C  25%                                          D  36%

3. *There are* 360 *students in a school.* 45% *are boys. How many girls are in the school?*

A  162                                          B  180

C  195                                          D  198

4. *The table shows the number of visitors to a park over two months. Which month had the greatest percent change from the previous month?*

| Month    | April | May | June |
|----------|-------|-----|------|
| Visitors | 400   | 500 | 550  |

A  April to May (25% increase)                 B  April to May (20% increase)

C  May to June (10% increase)                  D  May to June (25% increase)

5. *A recipe calls for* 2 *cups of flour. You used* 2.4 *cups. What is the percent error?*

A  0.4%                                         B  4%

C  16.7%                                        D  20%

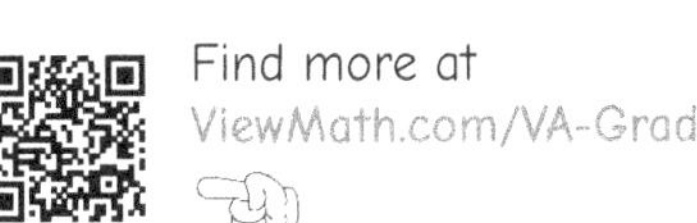
Find more at
ViewMath.com/VA-Grade7

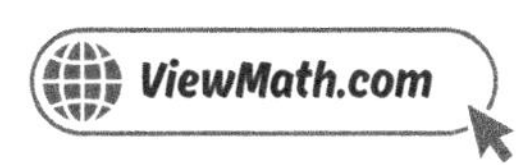

6. *Find the sum:* $5 + (-11) + 3 + (-2)$.

7. *Which expression is equivalent to* $(-5) - 3$?

   (A) $(-5) + 3$          (B) $(-5) + (-3)$

   (C) $5 + 3$              (D) $5 + (-3)$

8. *What must you add to* $-\dfrac{3}{8}$ *to get* $\dfrac{1}{2}$?

   (A) $\dfrac{1}{8}$             (B) $\dfrac{7}{8}$

   (C) $-\dfrac{1}{8}$         (D) $-\dfrac{7}{8}$

9. *Expand* $-5(x + 2)$.

   (A) $-5x + 10$         (B) $-5x + 2$

   (C) $-5x - 10$         (D) $5x - 10$

10. *A rectangle has area* $18x + 12$. *If the width is 6, what is the length?*

11. *Solve* $\dfrac{3x}{4} - 2 = 7$.

Get Online

Find more at
ViewMath.com/VA-Grade7

**ViewMath.com**

12. Solve $5x - 10 \geq 25$.

A   $x \geq 3$            B   $x \geq 7$

C   $x \leq 7$            D   $x \geq 35$

13. Solve $10 - 4x < 2$. Describe the graph.

14. A map has a scale of 1 cm = 50 km. Keisha measures the distance between two cities as 3.2 cm. What is the actual distance?

A   150 km            B   160 km

C   320 km            D   15.6 km

15. A floor plan uses 1 cm = 5 ft. You redraw at 1 cm = 2 ft. A room that is 3 cm by 4 cm on the original drawing. What is the area on the new drawing?

A   $12\ cm^2$            B   $30\ cm^2$

C   $75\ cm^2$            D   $150\ cm^2$

16. What shape is the cross-section when a cone is sliced vertically through its apex?

17. Two supplementary angles are equal. What is the measure of each?

A   45°            B   60°

C   90°            D   180°

Find more at
ViewMath.com/VA-Grade7

ViewMath.com

18. Point $C(5, -3)$ is reflected over the $y$-axis. What are the coordinates of $C'$?

   A) $(5, 3)$

   B) $(-5, -3)$

   C) $(-5, 3)$

   D) $(3, -5)$

19. Triangle $ABC$ has sides 3, 4, 5. Triangle $DEF$ is similar with the longest side equal to 20. What is the shortest side of triangle $DEF$?

   A) 8

   B) 12

   C) 15

   D) 16

20. A rectangular room is 14 ft by 10 ft. There is a built-in closet that is 3 ft by 4 ft. What is the area of the room not including the closet?

   A) 128 $ft^2$

   B) 140 $ft^2$

   C) 152 $ft^2$

   D) 12 $ft^2$

21. A cube has a surface area of $384$ $in^2$. What is the side length?

22. A rectangular fish tank is 50 cm long, 30 cm wide, and 40 cm tall. What is the volume?

   A) 120 $cm^3$

   B) 6,000 $cm^3$

   C) 60,000 $cm^3$

   D) 600 $cm^3$

Find more at
ViewMath.com/VA-Grade7

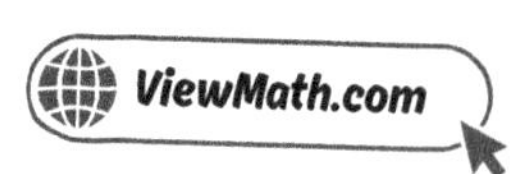

23. A cylindrical silo has a radius of 6 m and a height of 14 m. What is the volume? Use $\pi \approx 3.14$.

A  263.76 $m^3$                                         B  527.52 $m^3$

C  1,582.56 $m^3$                                       D  3,165.12 $m^3$

24. A school has 800 students. The principal surveys 50 students about their favorite subject. What is the population?

A  The 50 surveyed students                            B  All 800 students in the school

C  The principal                                       D  The students who like math

25. Team A's scores: mean 72, MAD 3. Team B's scores: mean 74, MAD 8. Which team is more consistent?

A  Team A because its MAD is smaller                   B  Team B because its mean is higher

C  Team A because its mean is lower                     D  Team B because its MAD is larger

26. Two groups have the same mean but different MADs. What does this tell you?

A  One group scored higher than the other              B  The groups have different levels of variability

C  The groups are identical                            D  The means were calculated incorrectly

27. Using the stem-and-leaf plot: Stem 7: 0 3 5 8; Stem 8: 1 4 6; Stem 9: 0 5. Find the median.

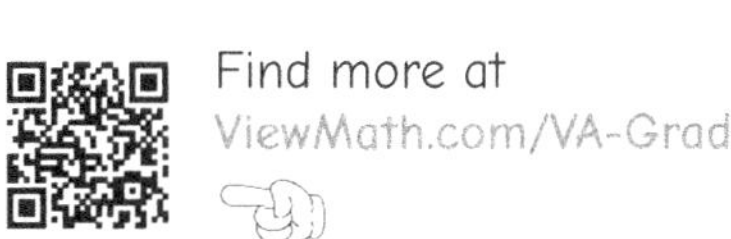

28. A spinner has 3 equal sections. After 90 spins, Section A appeared 35 times, Section B appeared 28 times, and Section C appeared 27 times. Based on this experiment, is the spinner likely fair?

A  No, because the results are not exactly equal

B  Yes, because the results are reasonably close to 30 each

C  No, because Section A appeared too many times

D  Yes, because 90 is divisible by 3

29. The table shows all possible products when two dice are rolled. How many outcomes give a product of 6? Write the probability as a fraction in simplest form.

| × | 1 | 2 | 3 | 4 | 5 | 6 |
|---|---|---|---|---|---|---|
| 1 | 1 | 2 | 3 | 4 | 5 | 6 |
| 2 | 2 | 4 | 6 | 8 | 10 | 12 |
| 3 | 3 | 6 | 9 | 12 | 15 | 18 |
| 4 | 4 | 8 | 12 | 16 | 20 | 24 |
| 5 | 5 | 10 | 15 | 20 | 25 | 30 |
| 6 | 6 | 12 | 18 | 24 | 30 | 36 |

30. Which of the following is the FIRST step in designing a simulation?

A  Record the results

B  Run many trials

C  Identify the event and its possible outcomes

D  Calculate the experimental probability

Get Online

Find more at
ViewMath.com/VA-Grade7

ViewMath.com

 # End of Practice Test 5 

*Great job finishing the test!*

 *My Score*

*I got _____________ out of 30 questions right.*

📊 *Check Your Score Online!*

*Visit **ViewMath Academy** to enter your answers and see which topics you need to review. You can also explore lessons, take quizzes, track your scores, and save your progress!*

*viewmath.com/score/7.1.VA.20*

# 6

# Practice Test 6

📋 30 Questions

---

## ✏️ Before You Start ✏️

- ✓ **Read each question carefully** before choosing your answer.
- ✓ **Show your work** on scratch paper when you need to.
- ✓ **Skip hard questions** and come back to them later.
- ✓ **Check your answers** when you're done.
- ✓ **Take your time** — there's no rush!

⭐ You've Got This! ⭐

Write an equation for the proportional relationship where 3 tickets cost $22.50.

What is 40% of 150?

| | |
|---|---|
| A  45 | B  50 |
| C  55 | D  60 |

3. Which equation is equivalent to the proportion $\dfrac{n}{250} = \dfrac{16}{100}$?

| | |
|---|---|
| A  $n = 250 \div 16$ | B  $n = 16 \times 250$ |
| C  $100n = 250 \times 16$ | D  $16n = 250 \times 100$ |

A bike costs $200. It is on sale for 35% off. What is the sale price?

| | |
|---|---|
| A  $70 | B  $130 |
| C  $135 | D  $165 |

Two students estimated the number of beans in a jar. The actual count was 200. Student A guessed 180; Student B guessed 220. Who had the greater percent error?

| | |
|---|---|
| A  Student A | B  Student B |
| C  Same percent error | D  Cannot be determined |

6. What is $9 + (-14)$?

| | |
|---|---|
| A  23 | B  $-23$ |
| C  $-5$ | D  5 |

Find more at
ViewMath.com/VA-Grade7

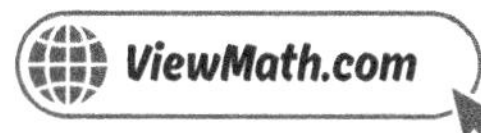

7. What is the distance between $-9$ and $4$ on a number line?

   A) 5 units

   B) $-5$ units

   C) 13 units

   D) $-13$ units

8. What is $\dfrac{3}{4} - 1\dfrac{1}{2}$?

   A) $\dfrac{3}{4}$

   B) $-\dfrac{3}{4}$

   C) $\dfrac{1}{4}$

   D) $-\dfrac{1}{4}$

9. Expand $-2(3x + 7)$.

   A) $-6x + 7$

   B) $-6x - 14$

   C) $-6x + 14$

   D) $6x - 14$

10. Factor $-9x - 6$.

    A) $-3(3x - 2)$

    B) $-3(3x + 2)$

    C) $3(-3x - 2)$

    D) $-9(x + 6)$

11. Solve $-0.2n + 5.6 = 4$

    A) $n = 8$

    B) $n = -8$

    C) $n = 48$

    D) $n = -48$

Find more at
ViewMath.com/VA-Grade7

**ViewMath.com**

12. *Which inequality is represented by the number line below?*

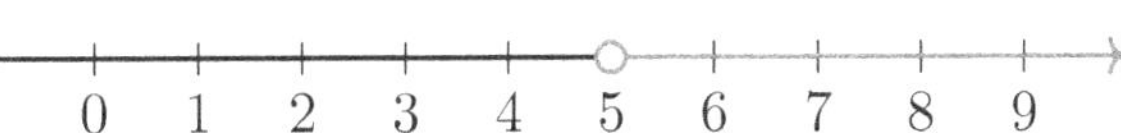

A  $x \geq 5$                                    B  $x > 5$

C  $x \leq 5$                                    D  $x < 5$


*Solve $7x - 3 \geq 18$. Give the smallest integer that is a solution.*


*A model car uses a scale of $1 : 24$. The real car is $180$ inches long. How long is the model?*

A  7.5 inches                              B  15 inches

C  24 inches                               D  156 inches


15. *A blueprint uses $1\ in = 12\ ft$. You redraw at $1\ in = 4\ ft$. A door that is $0.5\ in$ wide on the original becomes how wide?*

A  0.5 in                                  B  1 in

C  1.5 in                                  D  3 in


*A hexagonal prism is sliced parallel to its base. What is the cross-section?*

A  Rectangle                              B  Triangle

C  Hexagon                                D  Circle

Find more at
ViewMath.com/VA-Grade7

ViewMath.com

17. *Find the supplement of 43°.*

18. *Point $S(-3, -5)$ is reflected over the $x$-axis, then translated right 4. What are the final coordinates?*

19. *Two similar triangles have a scale factor of $\frac{2}{5}$. If a side of the larger triangle is 30 cm, what is the corresponding side of the smaller triangle?*

    (A) 6 cm                (B) 12 cm

    (C) 15 cm             (D) 75 cm

20. *A trapezoidal garden has parallel sides of 8 m and 12 m, and a height of 5 m. What is its area?*

21. *A rectangular prism is 6 ft by 6 ft by 10 ft. What is its surface area?*

    (A) $72\ ft^2$           (B) $240\ ft^2$

    (C) $312\ ft^2$          (D) $360\ ft^2$

22. *A triangular prism has a triangular base with base 10 m and height 6 m. The prism is 8 m long. What is the volume?*

Find more at
ViewMath.com/VA-Grade7

23. A cylindrical container has a volume of $628$ cm$^3$ and a height of 8 cm. What is the radius? Use $\pi \approx 3.14$.

A  3 cm

B  4 cm

C  5 cm

D  6 cm

24. A random sample means that:

A  Only the best members are chosen

B  The largest possible group is chosen

C  Every member of the population has an equal chance of being selected

D  Only volunteers are selected

25. When is a dot plot a better choice than a box plot for comparing two groups?

A  When data sets are very large

B  When you want to see every individual data point

C  When you only want to compare medians

D  When the data is categorical

26. Data set: $12, 15, 18, 20, 22, 25, 28$. What is the median?

A  18

B  20

C  22

D  19

27. A back-to-back stem-and-leaf plot is used to:

A  Show one data set with two stems

B  Compare two data sets using the same stems

C  Display data that has been doubled

D  Show data in a circle

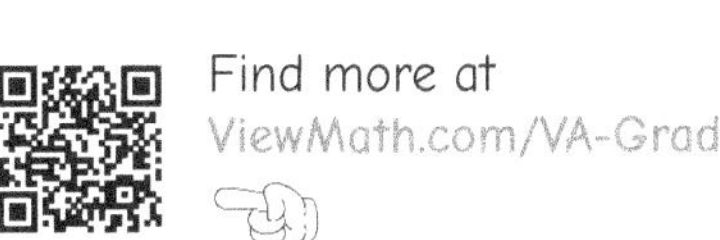
Find more at
ViewMath.com/VA-Grade7

28. A bag contains marbles of unknown colors. After $40$ draws (with replacement), red is drawn $14$ times. What is the experimental probability of drawing red?

   (A) $\frac{14}{40}$

   (B) $\frac{14}{26}$

   (C) $\frac{26}{40}$

   (D) $\frac{40}{14}$

29. A coin is flipped and a die is rolled. What is the probability of getting tails and an even number?

   (A) $\frac{1}{12}$

   (B) $\frac{1}{6}$

   (C) $\frac{1}{4}$

   (D) $\frac{1}{2}$

30. What is a simulation?

   (A) A way to calculate exact probabilities

   (B) A model that uses random numbers or objects to imitate a real event

   (C) A method that always gives the same result

   (D) A graph that shows all possible outcomes

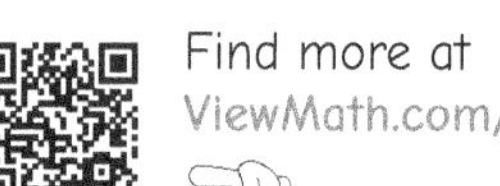

 # End of Practice Test 6 

*Great job finishing the test!*

 My Score

*I got _____________ out of 30 questions right.*

📊 Check Your Score Online!

*Visit **ViewMath Academy** to enter your answers and see which topics you need to review. You can also explore lessons, take quizzes, track your scores, and save your progress!*

viewmath.com/score/7.1.VA.21

# 7

# Practice Test 7

 30 Questions

## ✏️ Before You Start ✏️

- ✔ **Read each question carefully** before choosing your answer.
- ✔ **Show your work** on scratch paper when you need to.
- ✔ **Skip hard questions** and come back to them later.
- ✔ **Check your answers** when you're done.
- ✔ **Take your time** — there's no rush!

⭐ You've Got This! ⭐

Write an equation for the table and use it to find $y$ when $x = 20$.

| $x$ | $y$ |
|---|---|
| 4 | 18 |
| 6 | 27 |
| 8 | 36 |

36 is 45% of what number?

A  16.2

B  72

C  80

D  90

3.  48 is what percent of 320?

A jacket cost \$75 last winter. This winter it costs \$60. What is the percent decrease?

A  15%

B  20%

C  25%

D  30%

A baker expected to make 80 cookies from a recipe but made 72. What is the percent error (rounded to the nearest tenth)?


Get Online

Find more at
ViewMath.com/VA-Grade7

ViewMath.com

6. What is $(-8) + (-5)$?

    A) 13                                                  B) $-13$

    C) $-3$                                                 D) 3

7. An airplane is at 30,000 feet. It descends 8,500 feet. What is the new altitude?

8. What is $-\dfrac{7}{8} - \dfrac{1}{4}$?

    A) $-\dfrac{5}{8}$                                           B) $-\dfrac{9}{8}$

    C) $\dfrac{9}{8}$                                            D) $-\dfrac{3}{8}$

9. Expand $3(x + 4)$.

    A) $3x + 4$                                        B) $3x + 12$

    C) $3x + 7$                                        D) $12x$

10. Factor $16x + 24y - 8$.

    A) $4(4x + 6y - 2)$                         B) $8(2x + 3y - 1)$

    C) $2(8x + 12y - 4)$                      D) $8(2x + 3y + 1)$

11. A store marks down a jacket by $\dfrac{1}{4}$ of its original price $p$, then subtracts an additional \$8. The sale price is \$37. Find the original price.

Get Online

Find more at
ViewMath.com/VA-Grade7

ViewMath.com

12. A movie theater requires you to be more than 48 inches tall for a ride. You are 42 inches tall and grow $g$ inches per year. Which inequality finds how many years until you are tall enough?

A $42 + g > 48$

B $42g > 48$

C $48 - g > 42$

D $42 - g > 48$

13. Solve $4x + 3 \leq 19$ and describe the graph.

A Closed circle at 4, shade left

B Open circle at 4, shade left

C Closed circle at 4, shade right

D Closed circle at 5.5, shade left

14. Which statement about scale drawings is always true?

A Angles in the drawing are different from the actual angles

B The scale factor changes depending on which part of the drawing you measure

C All lengths in the drawing are multiplied by the same scale factor to get actual lengths

D If the scale factor is 4, the area of the actual figure is 4 times the drawing area

15. A drawing uses $1 \text{ cm} = 12 \text{ m}$. You redraw at $1 \text{ cm} = 4 \text{ m}$. A hallway is 5 cm on the original. How long is it on the new drawing?

16. You slice a cylinder with a horizontal cut parallel to its base. What shape is the cross-section?

A Rectangle

B Circle

C Oval

D Triangle

Find more at
ViewMath.com/VA-Grade7

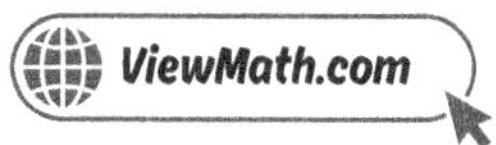

17. Two angles are supplementary. One angle is 115°. What is the other angle?

A) 25°

B) 55°

C) 65°

D) 75°

18. A point is at the origin $(0,0)$. It is rotated 90° clockwise. Where does it end up?

A) $(0,0)$

B) $(1,0)$

C) $(0,-1)$

D) $(0,1)$

19. A photo is 4 inches by 6 inches. It is enlarged so the longer side is 15 inches. What is the shorter side of the enlarged photo?

A) 8 in

B) 10 in

C) 12 in

D) 13 in

20. An H-shaped figure is made of three rectangles. The top and bottom bars are each 10 cm by 2 cm. The middle vertical bar is 2 cm by 6 cm. What is the total area?

21. A cereal box is 30 cm tall, 20 cm wide, and 6 cm deep. What is its surface area?

A) $1,440 \ cm^2$

B) $1,560 \ cm^2$

C) $1,800 \ cm^2$

D) $3,600 \ cm^2$

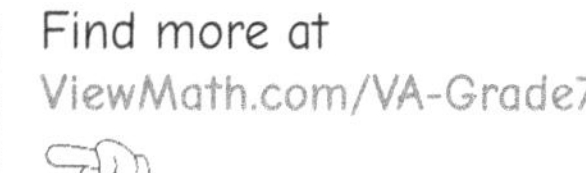
Find more at
ViewMath.com/VA-Grade7

22. A storage container is 2 ft long, 1.5 ft wide, and 3 ft tall. How many cubic feet can it hold?

A  $6 \; ft^3$

B  $6.5 \; ft^3$

C  $9 \; ft^3$

D  $10 \; ft^3$

23. Find the volume of a cylinder with radius 6 cm and height 10 cm. Use $\pi \approx 3.14$.

24. The diagram below shows four different sampling methods used to survey students at a school. Which method is most likely to produce an unbiased sample?

**Method A**

Survey the football team

**Method B**

Survey every 10th student on the school roster

**Method C**

Post survey online and wait for responses

**Method D**

Survey students in one math class

A  Method A

B  Method B

C  Method C

D  Method D

25. Class A: median 90, IQR 6. Class B: median 83, IQR 14. Which class is more consistent?

26. *Class A: median 85, IQR 8, range 20. Class B: median 80, IQR 15, range 40. A student says "Class B is better because it has a wider range." Is the student correct?*

A. *Yes — wider range means more talent*

B. *No — wider range means less consistency, and Class A has a higher median*

C. *Yes — Class B's scores spread more, which is always better*

D. *No — range does not measure anything useful*

27. *Create a stem-and-leaf plot for the following test scores and find the mean:*

$85, 78, 92, 81, 88, 75, 90, 83, 79, 87$

| Stem | Leaf |
|------|------|
| 7 | |
| 8 | |
| 9 | |

*Key:* $7 \mid 5$ *means* $75$

28. *A coin is flipped 200 times. Heads comes up 112 times. What is the experimental probability of heads?*

A. $0.50$

B. $0.56$

C. $0.44$

D. $0.88$

29. *A coin is flipped and a die is rolled. What is the probability of getting heads and a number greater than 4?*

A. $\frac{1}{12}$

B. $\frac{1}{6}$

C. $\frac{2}{12}$

D. $\frac{1}{3}$

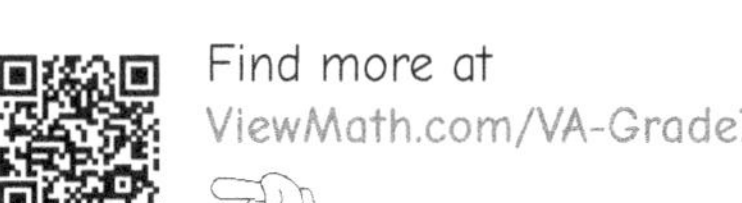

30. *A game has a 25% chance of winning. Which simulation model correctly represents this?*

A  *Flip a coin — heads means win*

B  *Roll a die — roll a 1 means win*

C  *Use a spinner with 4 equal sections — one section means win*

D  *Use digits 0–9 — digits 0–4 mean win*

Find more at
ViewMath.com/VA-Grade7

# ⭐ End of Practice Test 7 ⭐

*Great job finishing the test!*

 My Score

*I got _____________ out of 30 questions right.*

📊 Check Your Score Online!

Visit **ViewMath Academy** to enter your answers and see which topics
you need to review. You can also explore lessons, take quizzes, track
your scores, and save your progress!

*viewmath.com/score/7.1.VA.22*

# 8

# Practice Test 8

 30 Questions

---

## ✏️ Before You Start ✏️

- ✓ **Read each question carefully** before choosing your answer.
- ✓ **Show your work** on scratch paper when you need to.
- ✓ **Skip hard questions** and come back to them later.
- ✓ **Check your answers** when you're done.
- ✓ **Take your time** — there's no rush!

 You've Got This! 

Do your best and show what you know!

If $y = 6x$, what is $y$ when $x = 7$?

A   13

B   36

C   42

D   48

The circle graph shows how 200 students get to school. How many students take the bus?

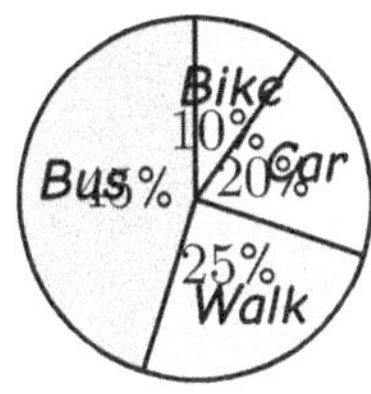


3. 91 is 65% of what number?

A gym membership increased from $45 per month to $54 per month. What is the percent increase?

A   9%

B   15%

C   18%

D   20%

A lab experiment expected to produce 250 mL but produced 240 mL. Find the percent error.

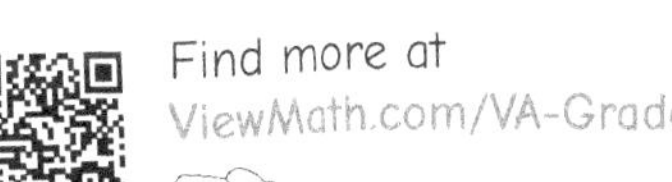
Find more at
ViewMath.com/VA-Grade7

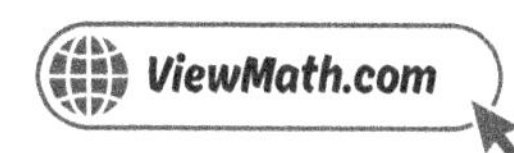

6. Look at the table of deposits and withdrawals from a bank account.

| Day | Transaction |
| --- | --- |
| Monday | +$30 |
| Tuesday | −$45 |
| Wednesday | +$20 |
| Thursday | −$15 |

If the account started at $0, what is the balance after Thursday?

A  $110

B  −$10

C  $10

D  −$50

7. Which subtraction problem gives a positive result?

A  $(-6) - 2$

B  $3 - 11$

C  $(-1) - (-9)$

D  $(-5) - 3$

8. The temperature is −6.4°C. It rises 9.2°C. What is the new temperature?

A  $-15.6°C$

B  $15.6°C$

C  $-2.8°C$

D  $2.8°C$

9. Expand and simplify $2(x + 3) + 5$.

A  $2x + 8$

B  $2x + 11$

C  $2x + 10$

D  $7x + 3$

10. Show that $4(5x + 3)$ and $20x + 12$ are equivalent by expanding.

11. Use the fraction bar model to solve: $\dfrac{2}{3}x = 8$. What is $x$?

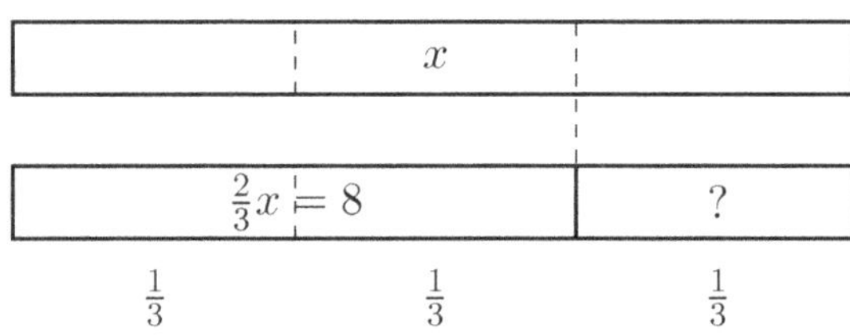


A  $x = 12$

B  $x = 16$

C  $x = 10$

D  $x = 5\dfrac{1}{3}$

12. Solve $3x + 7 \le 25$.

A  $x \le \dfrac{32}{3}$

B  $x \le 6$

C  $x \ge 6$

D  $x \le 9$

13. What inequality is shown by a closed circle at $-6$ with shading to the right?

14. A scale drawing of a soccer field is shown on the grid below. Each grid square represents $10\ m \times 10\ m$ in real life. What is the actual perimeter of the field?

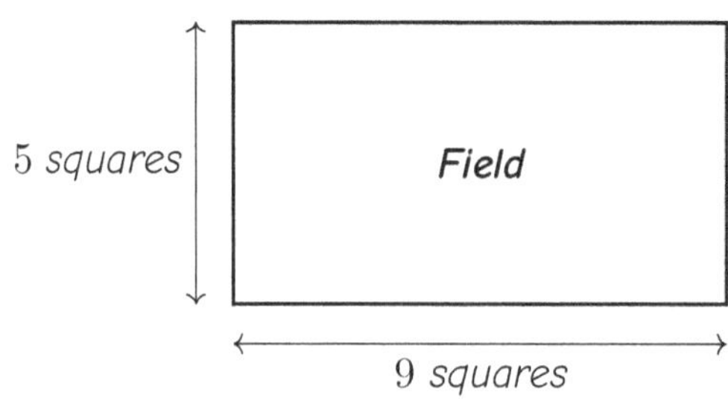


15. A floor plan uses $1\ cm = 6\ ft$. On this plan, a room is $9\ cm$ long. On a second plan, the same room is $3\ cm$ long. What scale does the second plan use?

16. A square pyramid has a base edge of 8 cm. It is sliced parallel to the base halfway up. Is the cross-section larger, smaller, or the same size as the base?

17. Two vertical angles are $(7x + 1)°$ and $(5x + 13)°$. Find $x$ and the angle measure.

Get Online

Find more at
ViewMath.com/VA-Grade7

18. Which of the following is true about all three rigid transformations (translation, reflection, rotation)?

A  They all change the size of the figure

B  They all preserve the size and shape of the figure

C  They all keep the figure's orientation the same

D  They all move the figure to the right

19. Rectangle $PQRS$ is similar to Rectangle $WXYZ$. $PQ = 5$, $QR = 8$, $WX = 12.5$. Find $XY$.

20. The diagram shows a figure made of a rectangle and a triangle. What is the total area?

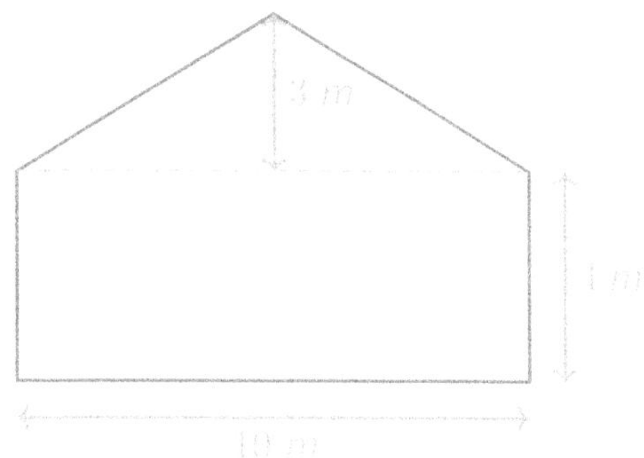


A  40 $m^2$

B  55 $m^2$

C  70 $m^2$

D  85 $m^2$

21. A shipping box is 12 in by 8 in by 6 in. What is the total surface area?

22. Which formula is used to find the volume of any prism?

A  $V = lwh$ only

B  $V = Bh$

C  $V = 2(lw + lh + wh)$

D  $V = \frac{1}{3}Bh$

Get Online

Find more at
ViewMath.com/VA-Grade7

ViewMath.com

23. A cylinder has a diameter of 8 cm and a height of 5 cm. What is its volume? Use $\pi \approx 3.14$.

A $100.48\ cm^3$

B $200.96\ cm^3$

C $251.2\ cm^3$

D $401.92\ cm^3$

24. A random sample of 50 apples from an orchard found that 6 were bruised. If the orchard has 3,000 apples, about how many are bruised?

25. Team A: mean 60, MAD 4. Team B: mean 70, MAD 3. Is the difference in means meaningful? Explain briefly.

26. Data set: $20, 22, 25, 27, 30, 32, 35$. What is the IQR?

A 15

B 10

C 7

D 5

27. How many data values are shown in this stem-and-leaf plot?

Stem 5: 1 3 5     Stem 6: 0 2 4 7 8     Stem 7: 1 6     Stem 8: 0

28. Tyler draws a card from a standard deck, records the suit, and replaces it. He does this 52 times. Hearts appear 16 times. What is the experimental probability of drawing a heart?

A $\frac{13}{52}$

B $\frac{16}{52}$

C $\frac{16}{36}$

D $\frac{4}{13}$

Find more at
ViewMath.com/VA-Grade7

A coin is flipped and a die is rolled. What is the probability of getting tails and a 1? Write your answer as a fraction.

30. How does increasing the number of trials in a simulation affect the results?

A  The results become less reliable

B  The experimental probability gets closer to the theoretical probability

C  The results stay exactly the same

D  The simulation takes less time

 # End of Practice Test 8

*Great job finishing the test!*

 ## My Score

I got _____________ out of 30 questions right.

Check your answers in the Answer Key at the back of the book.

Review any questions you missed. That's how we learn!

## Check Your Score Online!

Visit **ViewMath Academy** to enter your answers and see which topics you need to review. You can also explore lessons, take quizzes, track your scores, and save your progress!

viewmath.com/score/7.1.VA.23

Or go to viewmath.com/score and enter code: 7.1.VA.23

# 9

# Practice Test 9

☑ 30 Questions

## ✎ Before You Start ✎

- ✓ **Read each question carefully** before choosing your answer.
- ✓ **Show your work** on scratch paper when you need to.
- ✓ **Skip hard questions** and come back to them later.
- ✓ **Check your answers** when you're done.
- ✓ **Take your time** — there's no rush!

★ You've Got This! ★

1. *The point $(8, 52)$ is on a proportional graph. Which equation describes the relationship?*

A  $y = 6x$                                    B  $y = 6.5x$

C  $y = 7x$                                    D  $y = 8x$


2. *What is 25% of 80?*

A  15                                          B  20

C  25                                          D  32


3. Solve using a proportion: 72 is 90% of what number?

A  64.8                                        B  75

C  78                                          D  80


4. *What multiplier represents a 40% decrease?*

A  0.40                                        B  0.60

C  1.40                                        D  1.60


5. *If your estimate is exactly correct, what is the percent error?*

A  0%                                          B  1%

C  50%                                         D  100%


6. *What is $(-3) + (-9) + 4$?*

A  $-8$                                        B  $-16$

C  10                                          D  2


Get Online

Find more at
ViewMath.com/VA-Grade7

ViewMath.com

7. A scuba diver is at $-25$ feet. She swims up 10 feet then descends 7 feet. What is her new depth?

(A) $-22$ feet

(B) $-32$ feet

(C) $-42$ feet

(D) $-12$ feet

8. A thermometer shows these temperature readings over three recordings.

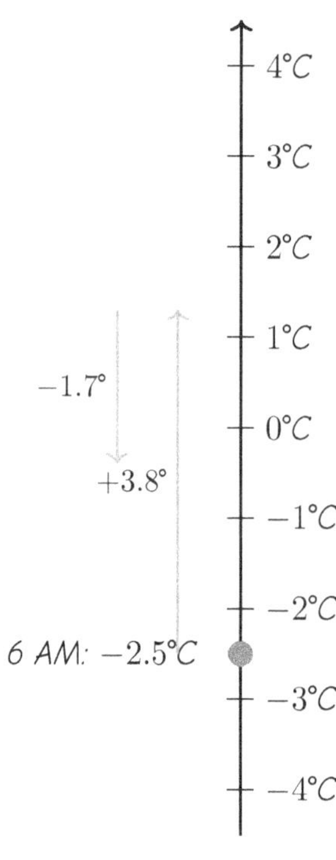


What is the final temperature reading?

9. A student expands $-3(x - 4)$ and writes $-3x - 12$. What mistake did the student make?

(A) Forgot to distribute to the second term

(B) Did not change the sign when multiplying $-3 \times (-4)$

(C) Multiplied only the first term

(D) Added instead of multiplying

Get Online

Find more at
ViewMath.com/VA-Grade7

ViewMath.com

The total area of the two rectangles shown below is $8x + 12$. Both rectangles have the same height $h$. Find the value of $h$.

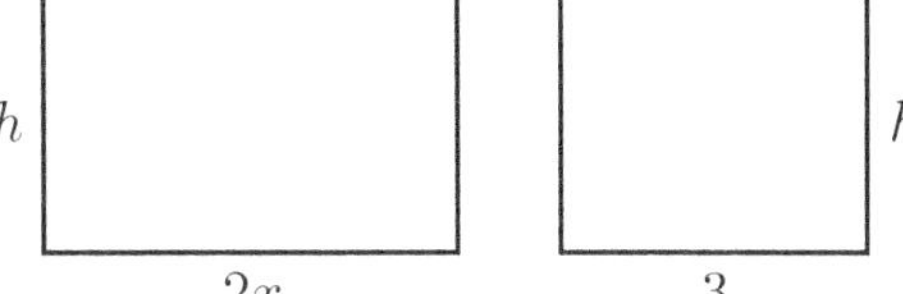



Solve $\dfrac{x}{6} - \dfrac{2}{3} = \dfrac{1}{2}$.

A  $x = 3$                                              B  $x = 7$

C  $x = -1$                                       D  $x = 10$

12. Solve $2x + 3 > 11$.

A  $x > 4$                                             B  $x > 7$

C  $x < 4$                                           D  $x > 14$

True or false: The graph of $x < 5$ and $x \le 5$ look exactly the same.

A  True — both shade to the left of 5         B  False — $x < 5$ has an open circle, $x \le 5$ has
a closed circle

C  False — they shade in different directions      D  True — the circle type does not matter

A blueprint uses the scale $1\ in = 6\ ft$. A hallway is $3.5\ in$ long on the blueprint. What is the actual length?

A  18 ft                                              B  21 ft

C  24 ft                                              D  9.5 ft

Get Online

Find more at
ViewMath.com/VA-Grade7

ViewMath.com

15. *A model uses scale $1 : 25$. A new model uses scale $1 : 75$. A window that is 6 cm on the first model becomes how long on the second?*

A) 2 cm

B) 6 cm

C) 18 cm

D) 150 cm

16. *A cube can be sliced to produce a hexagonal cross-section. How?*

A) By cutting parallel to the base

B) By cutting diagonally through all six faces

C) By cutting vertically through the center

D) It is impossible to get a hexagon from a cube

17. *Two vertical angles are $(5x - 20)°$ and $(3x + 10)°$. What is the measure of each angle?*

A) 45°

B) 55°

C) 60°

D) 75°

18. *Point $D(2, 6)$ is rotated 90° clockwise about the origin. What are the coordinates of $D'$?*

A) $(-6, 2)$

B) $(6, -2)$

C) $(-2, -6)$

D) $(6, 2)$

19. *Two similar triangles have corresponding sides of 7 and 21. What is the ratio of their areas?*

A) $1 : 3$

B) $1 : 7$

C) $1 : 9$

D) $7 : 21$

Find more at
ViewMath.com/VA-Grade7

20. A shape is made of a rectangle 8 in by 6 in and two semicircles, each with a diameter of 6 in, attached to both short sides. What is the total area? Use $\pi \approx 3.14$.

A. $48\ in^2$

B. $76.26\ in^2$

C. $104.52\ in^2$

D. $133.04\ in^2$

21. A rectangular prism has a surface area of $94\ cm^2$. Its length is 5 cm and width is 3 cm. What is its height?

A. $2\ cm$

B. $3\ cm$

C. $4\ cm$

D. $5\ cm$

22. What is the volume of a rectangular prism with length 6 cm, width 4 cm, and height 3 cm?

A. $13\ cm^3$

B. $24\ cm^3$

C. $48\ cm^3$

D. $72\ cm^3$

23. A cylinder has a diameter of 10 in and a height of 6 in. What is its volume? Use $\pi \approx 3.14$.

A. $188.4\ in^3$

B. $471\ in^3$

C. $942\ in^3$

D. $1,884\ in^3$

24. A researcher wants to know the average height of seventh graders in a state. She measures the heights of 100 seventh graders from different schools. What is the sample?

A. All seventh graders in the state

B. All students in the schools she visited

C. The 100 seventh graders she measured

D. The tallest seventh graders

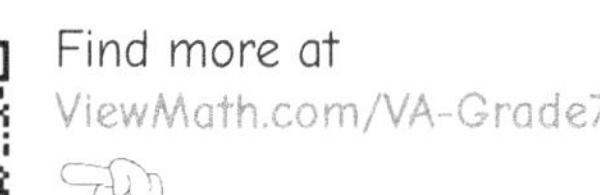
Find more at
ViewMath.com/VA-Grade7

25. The box plots below compare test scores for Class A and Class B.

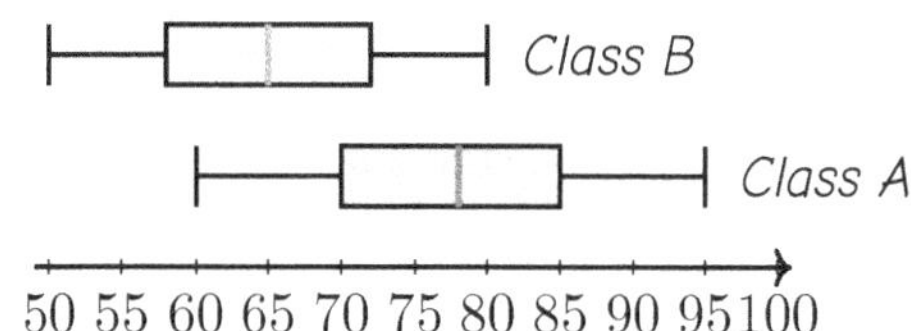



Which statement is best supported by the box plots?

(A) Class B scored higher than Class A

(B) Class A has a higher median and a larger IQR

(C) Class A has a higher median and both classes have the same IQR

(D) Class A and Class B overlap, but Class A's median is higher

26. Group A: $\{60, 65, 70, 75, 80\}$. Group B: $\{40, 55, 70, 85, 100\}$. Both have a mean of 70. Which group is more variable?

(A) Group A

(B) Group B

(C) They are equally variable

(D) Cannot be determined

27. The back-to-back stem-and-leaf plot compares quiz scores for two classes.

| Class A | Stem | Class B |
|---:|:---:|:---|
| 5 2 | 6 | 3 5 8 |
| 8 5 3 0 | 7 | 2 4 9 |
| 6 4 | 8 | 1 3 5 7 |
| 2 | 9 | 0 5 |

Key: 5 | 7 means 75 (Class A) and 7 | 2 means 72 (Class B)

Which class has a higher median?

(A) Class A

(B) Class B

(C) They have the same median

(D) Cannot be determined

A spinner with 4 equal sections is spun 80 times. Section A appears 25 times. How does the experimental probability compare to the theoretical probability?

A. Experimental (0.3125) is higher than theoretical (0.25)

B. Experimental (0.25) equals theoretical (0.25)

C. Experimental (0.20) is lower than theoretical (0.25)

D. Experimental (0.3125) is lower than theoretical (0.50)

Three coins are flipped. What is the probability of getting NO heads (all tails)?

A. $\frac{1}{2}$

B. $\frac{1}{4}$

C. $\frac{1}{8}$

D. $\frac{3}{8}$

30. Describe how to use a standard die to simulate a $\frac{1}{3}$ probability event.

Get Online

Find more at
ViewMath.com/VA-Grade7

# ⭐ *End of Practice Test 9* ⭐

*Great job finishing the test!*

###  *My Score*

*I got __________ out of 30 questions right.*

### 📊 *Check Your Score Online!*

*Visit **ViewMath Academy** to enter your answers and see which topics you need to review. You can also explore lessons, take quizzes, track your scores, and save your progress!*

*viewmath.com/score/7.1.VA.24*

# 10

# Practice Test 10

 30 Questions

 **Before You Start** 

- ✓ **Read each question carefully** before choosing your answer.
- ✓ **Show your work** on scratch paper when you need to.
- ✓ **Skip hard questions** and come back to them later.
- ✓ **Check your answers** when you're done.
- ✓ **Take your time** — there's no rush!

⭐ You've Got This! ⭐

Do your best and show what you know!

*The graph below shows the cost of buying pounds of apples. Which equation matches this graph?*

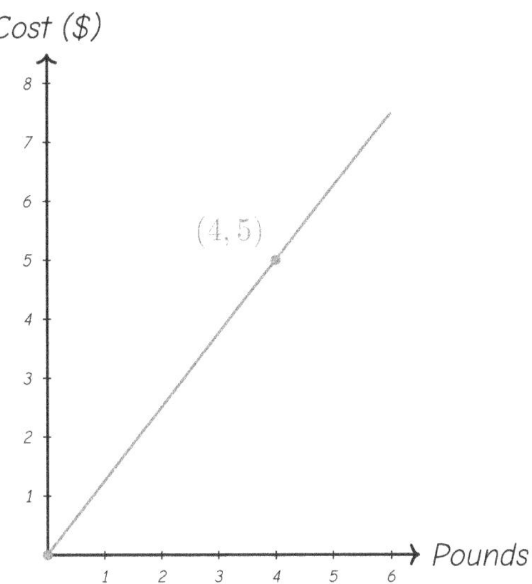


A  $c = 4p$                              B  $c = 1.25p$

C  $c = 5p$                              D  $c = 0.8p$

*The $10 \times 10$ grid below has some squares shaded. What percent of the grid is shaded?*

A  25%                                   B  30%

C  35%                                   D  40%

Get Online
Find more at
ViewMath.com/VA-Grade7

ViewMath.com

3. *Use a proportion to find 35% of 120.*

- A) 35
- B) 38
- C) 42
- D) 48

4. *Which of the following results in the greatest final value, starting from $100?*

- A) Increase by 30%
- B) Increase by 20%, then increase by 10%
- C) Increase by 15% twice
- D) All give the same result

5. *A package was estimated to weigh 10 kg, but it actually weighs 12 kg. A second package was estimated to weigh 50 kg, but it actually weighs 52 kg. Which estimate has the smaller percent error?*

- A) The first package
- B) The second package
- C) They have the same percent error
- D) Cannot be determined

6. *A thermometer shows temperature changes over three hours.*

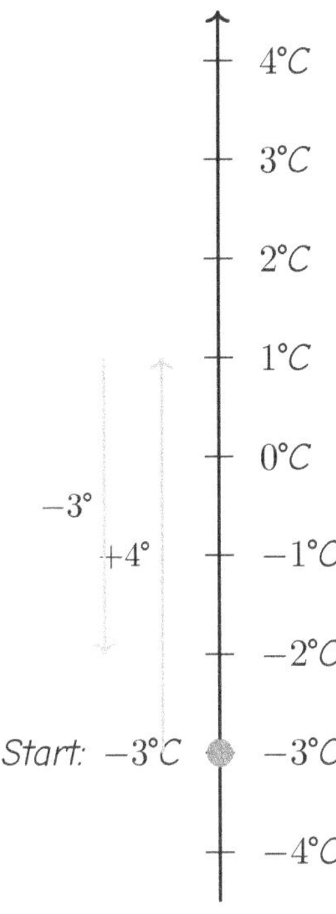


*What is the final temperature after these two changes?*

7. *What is $(-8) - (-15)$?*

8. *A hiker descends from an elevation of −15.4 meters to −23.8 meters. How many meters did the hiker descend?*


Get Online

Find more at
ViewMath.com/VA-Grade7

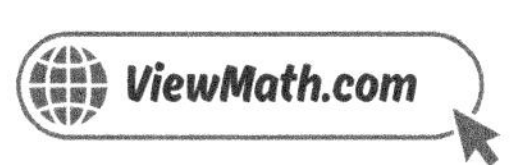
ViewMath.com

9. Expand $-4(2y - 5)$.

   (A) $-8y - 20$                    (B) $-8y - 5$

   (C) $-8y + 20$                    (D) $8y + 20$

10. Which is the completely factored form of $8m + 20$?

   (A) $2(4m + 10)$                (B) $4(2m + 5)$

   (C) $8(m + 20)$                (D) $4(2m + 20)$

11. The bar graph shows how much three students earned. Together they earned a total of $31.50. Student C earned $\frac{1}{2}$ as much as Student A. Find how much Student C earned.

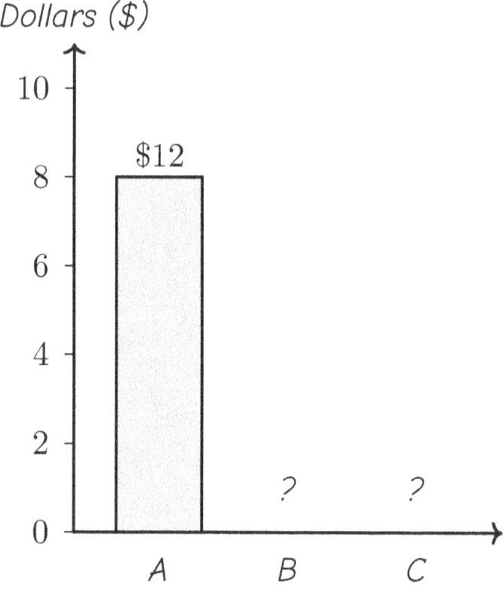


Student A earned $12, and Student C earned $\frac{1}{2}$ of Student A's earnings.

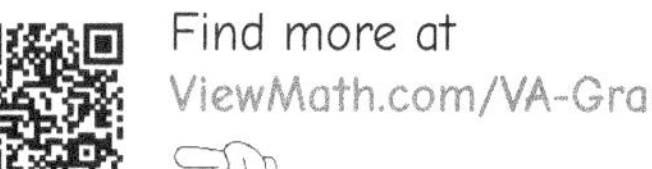

12. Solve $7 - 2x < 1$.

A  $x < 3$

B  $x > 3$

C  $x < -3$

D  $x > -3$


13. Solve $-3x + 9 > 0$. Then graph the solution on the number line below by describing the circle type and shading direction.

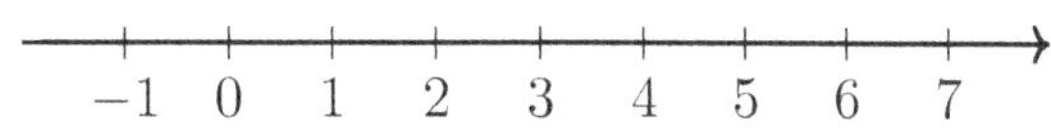

14. A rectangular swimming pool is 6 cm by 2.5 cm on a drawing with scale 1 cm = 8 m. What is the actual perimeter of the pool?

A  17 m

B  68 m

C  136 m

D  960 m


15. A drawing uses 1 cm = 5 km. A road is 7 cm long. Another map shows the same road as 3.5 cm. What is the scale of the second map?

A  1 cm = 2.5 km

B  1 cm = 5 km

C  1 cm = 10 km

D  1 cm = 17.5 km


16. What shape is the cross-section when you cut a sphere through its center?


Get Online

Find more at
ViewMath.com/VA-Grade7

ViewMath.com

17. Two lines intersect. One angle formed is 130°. What are the measures of all four angles?

  A) 130°, 130°, 50°, 50°
  B) 130°, 50°, 130°, 100°

  C) 130°, 130°, 130°, 130°
  D) 130°, 60°, 130°, 40°

18. The coordinate plane shows a rectangle with vertices $A(1,1)$, $B(5,1)$, $C(5,3)$, $D(1,3)$. Reflect the rectangle over the $x$-axis and list the new vertices.

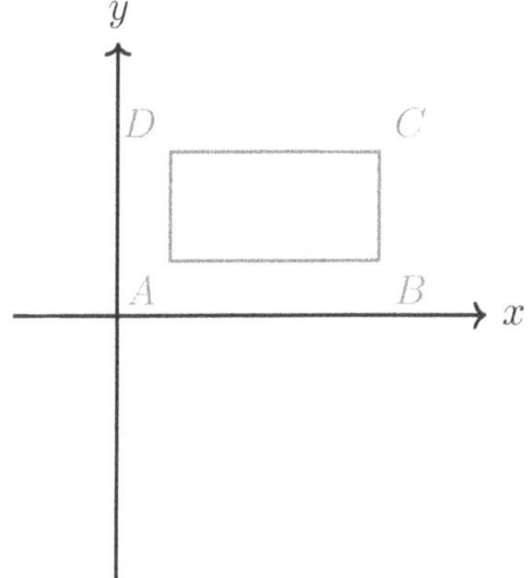


Your Answer

19. A building is 60 ft tall and casts a 40-ft shadow. A nearby tree casts a 12-ft shadow. How tall is the tree?

  A) 8 ft
  B) 12 ft

  C) 18 ft
  D) 20 ft

20. A rectangular yard is 20 m by 15 m. A circular flower bed with radius 3 m is in the center. What is the area of the yard NOT covered by the flower bed? Use $\pi \approx 3.14$.

  A) $271.74\ m^2$
  B) $300\ m^2$

  C) $328.26\ m^2$
  D) $28.26\ m^2$

Get Online

Find more at
ViewMath.com/VA-Grade7

ViewMath.com

21. What is the surface area of a cube with side length 4 cm?

   A   16 $cm^2$                                           B   64 $cm^2$

   C   96 $cm^2$                                           D   24 $cm^2$


22. A triangular prism has a triangular base with legs 5 cm and 12 cm (right triangle). The prism is 15 cm long. What is the volume?

   A   150 $cm^3$                                          B   300 $cm^3$

   C   450 $cm^3$                                          D   900 $cm^3$


23. A cylindrical swimming pool has a radius of 5 m and a depth of 2 m. How many cubic meters of water does it hold? Use $\pi \approx 3.14$.


24. The table below shows the results of two samples taken from a population of 1,000 students.

|                | Sample 1 (50 students) | Sample 2 (50 students) |
|----------------|------------------------|------------------------|
| Prefer Math    | 18                     | 22                     |
| Prefer Science | 14                     | 12                     |
| Prefer English | 10                     | 8                      |
| Prefer Art     | 8                      | 8                      |

Based on both samples, the best prediction for how many of the 1,000 students prefer math is:

   A   180                                                 B   220

   C   400                                                 D   360

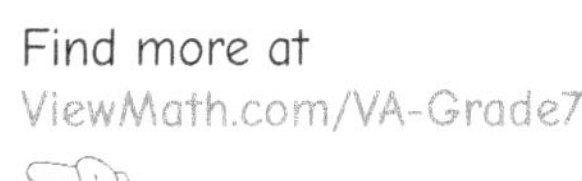
Find more at
ViewMath.com/VA-Grade7

25. Two histograms show that Group A's data is skewed left and Group B's data is symmetric. Which measure of center is most appropriate to compare them?

A  Mean for both

B  Median for both

C  Mode for both

D  Range for both

26. Data: $100, 200, 300, 400, 500$. Find the median and the IQR.

27. Which of the following is true about the leaves in a stem-and-leaf plot?

A  Leaves are listed in random order

B  Leaves are listed in order from least to greatest

C  Each leaf represents the tens digit

D  Leaves are always single digits from $1$ to $9$

The bar graph shows the results of rolling a die 60 times. Based on the graph, what is the experimental probability of rolling a 3?

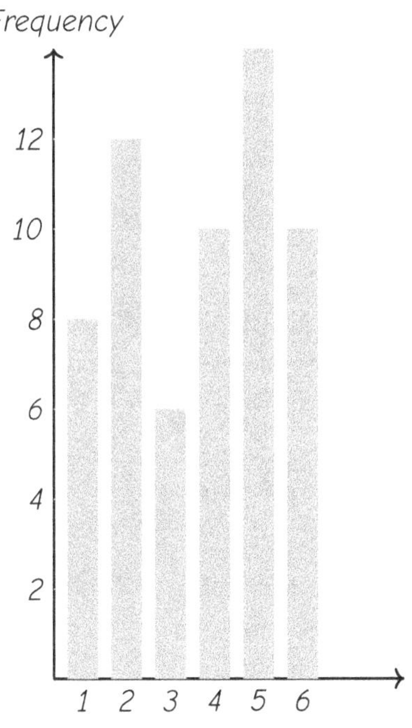


A   $\frac{6}{60}$

B   $\frac{10}{60}$

C   $\frac{12}{60}$

D   $\frac{14}{60}$

Two dice are rolled. What is the probability that the product (not the sum) of the two numbers is 12?

A   $\frac{2}{36}$

B   $\frac{4}{36}$

C   $\frac{3}{36}$

D   $\frac{6}{36}$

30. A student simulates rolling two dice 36 times to see how often the sum is 7. They get a sum of 7 a total of 8 times. How does this compare to the expected number?

A   Expected: 6; the simulation result (8) is slightly higher

B   Expected: 6; the simulation result (8) is much higher

C   Expected: 12; the simulation result (8) is lower

D   Expected: 7; the simulation result (8) is exact

 # End of Practice Test 10 

*Great job finishing the test!*

 **My Score**

I got _____________ out of 30 questions right.

Check your answers in the Answer Key in the back of the book.

Review any questions you missed. That's how we learn!

##  Check Your Score Online!

Visit **ViewMath Academy** to enter your answers and see which topics you need to review. You can also explore lessons, take quizzes, track your scores, and save your progress!

viewmath.com/score/7.1.VA.25

Or go to viewmath.com/score and enter code: 7.1.VA.25

# Answer Key & Explanations

# Answer Key

*First try each test on your own, then check your work here.*

## Practice Test 1 — Answer Key

**1** 40 beads   **2** C   **3** B   **4** D   **5** C   **6** 9   **7** C   **8** B   **9** $4x - 15$

**10** $-5(2m + 3)$   **11** A   **12** $9s + 35 \leq 200;\ s \leq 18$

**13** $85 - 7h < 20;\ h > 9\frac{2}{7}$ hours (about 9.3 hours)   **14** C   **15** 5 cm   **16** C   **17** 100° and 80°

**18** $Q'(-3, -4)$   **19** C   **20** B   **21** A   **22** B   **23** C   **24** B   **25** C   **26** B

**27** C   **28** $\frac{5}{36}$   **29** B   **30** 48

## 💡 Time to Learn! 💡

*Review the explanations below, **especially for the questions you missed**.*

*Understanding why each answer is correct builds stronger problem-solving skills.*

***Tip:*** *Circle any questions you got wrong, then read their explanation carefully.*

## Practice Test 1 — Detailed Explanations

**1** $6 = 0.15b$, so $b = 6 \div 0.15 = 40$ beads.

Find more at
ViewMath.com/VA-Grade7

Total students: $12 + 24 + 18 + 6 = 60$. Soccer: 24. Percent: $24 \div 60 = 0.40 = 40\%$.

$\frac{n}{80} = \frac{37.5}{100}$. Cross-multiply: $100n = 3000$. Divide: $n = 30$.

Change: $14 - 8 = 6$. Percent increase: $6 \div 8 = 0.75 = 75\%$.

$\frac{|35 - 10|}{40} \times 100 = \frac{5}{40} \times 100 = 12.5\%$.

$(-15) + n = -6$, so $n = -6 - (-15) = -6 + 15 = 9$.

Difference from B to A: $300 - (-200) = 300 + 200 = 500$ ft.

$5.6 - 8.1 = 5.6 + (-8.1) = -2.5$. The larger absolute value (8.1) is negative.

Expand: $-2x - 12 + 6x - 3$. Combine: $(-2x + 6x) + (-12 - 3) = 4x - 15$.

GCF of 10 and 15 is 5. Factor out $-5$: $-10m \div (-5) = 2m$ and $-15 \div (-5) = 3$. Result: $-5(2m + 3)$.

Equation: $-3.5 + 4t = 8.5$. Add 3.5: $4t = 12$. Divide by 4: $t = 3$.

Total cost: $9s + 35 \leq 200$. Subtract 35: $9s \leq 165$. Divide by 9: $s \leq 18.\overline{3}$. Since $s$ must be a whole number, $s \leq 18$.

$85 - 7h < 20$. Subtract 85: $-7h < -65$. Divide by $-7$ and flip: $h > \frac{65}{7} = 9\frac{2}{7}$.

Multiply the map distance by the scale factor: $6 \times 25 = 150$ km.

The scale ratio is $\frac{7}{7} = 1$. The length stays the same: $5 \times 1 = 5$ cm.

Find more at
ViewMath.com/VA-Grade7

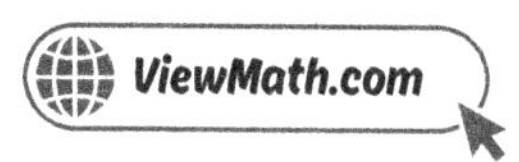

Every cross-section of a sphere is a circle. The size of the circle depends on where you cut, but the shape is always a circle.

$5x + 4x = 180$. Combine: $9x = 180$. Solve: $x = 20$. Angles: $5(20) = 100°$ and $4(20) = 80°$.

90° clockwise: $(x, y) \rightarrow (y, -x)$. $(4, -3) \rightarrow (-3, -4)$.

All circles have the same shape. Any circle can be scaled (dilated) to match any other circle. Therefore, all circles are similar to each other.

Rectangle: $6 \times 4 = 24$ cm². Triangle: $\frac{1}{2} \times 6 \times 2 = 6$ cm². Total: $24 + 6 = 30$ cm².

Two triangles: $2 \times \frac{1}{2} \times 6 \times 4 = 24$ cm². Three rectangles: $6 \times 10 + 5 \times 10 + 5 \times 10 = 60 + 50 + 50 = 160$ cm². Total: $24 + 160 = 184$ cm².

$s^3 = 64$. $s = 4$ cm because $4^3 = 64$.

$SA = 2\pi r^2 + 2\pi rh = 2(3.14)(25) + 2(3.14)(5)(8) = 157 + 251.2 = 408.2$ cm².

$\frac{15}{60} = \frac{1}{4} = 25\%$. So 25% of $600 = 150$.

A box plot shows the quartiles ($Q_1$ and $Q_3$), so you can read the IQR directly as $Q_3 - Q_1$.

Company X has smaller IQR (1.5 vs. 4) and range (4 vs. 8), meaning delivery times are more predictable.

Stem 8 with leaf 0 means 80.

$P = \frac{25}{180} = \frac{5}{36}$.

Get Online

Find more at
ViewMath.com/VA-Grade7

ViewMath.com

$P(spinner = 3) = \frac{1}{4}$, $P(die = 3) = \frac{1}{6}$. $P = \frac{1}{4} \times \frac{1}{6} = \frac{1}{24}$.

$80 \times 0.60 = 48$.

## ☑ Practice Test 2 — Answer Key

B      28%      120      C      B      A      D      $-\dfrac{1}{3}$      C

$12(3a + 2b)$      B      $x \leq -5$      $s \geq 87$; closed circle at 87, shade right

The area does not change      B      A      C      B      12.8 ft      B

C      $9\ ft^3$      B      4%      B

City A: mean $= 66.6°F$, range $= 10°F$. City B: mean $= 68.2°F$, range $= 18°F$. City A is more consistent.

C      A      C      B

## 💡 Time to Learn! 💡

Review the explanations below, **especially for the questions you missed.**

Understanding why each answer is correct builds stronger problem-solving skills.

**Tip:** Circle any questions you got wrong, then read their explanation carefully.

## 📖 Practice Test 2 — Detailed Explanations

$k = \frac{11}{2} = 5.5$. Equation: $y = 5.5x$. When $x = 9$: $y = 5.5 \times 9 = 49.5$.

Find more at
ViewMath.com/VA-Grade7

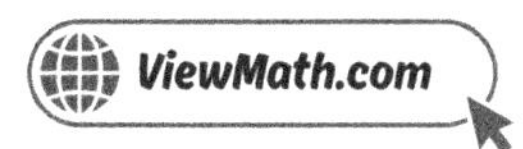

Shaded squares: 7 full rows of $10 = 70$, plus $2 = 72$. Unshaded: $100 - 72 = 28$. So 28% is not shaded.

$\frac{84}{w} = \frac{70}{100}$. Cross-multiply: $70w = 8400$. Divide: $w = 120$.

A 25% increase means you keep 100% and add 25%: $1 + 0.25 = 1.25$.

A 0% percent error means the estimated value equals the actual value exactly.

$(-7) + 7 = 0$ (opposites cancel). Then $0 + (-3) = -3$.

Subtracting a negative means adding: $(-7) + 12 = 5$.

$-\frac{5}{6} + \frac{1}{2} = -\frac{5}{6} + \frac{3}{6} = -\frac{2}{6} = -\frac{1}{3}$.

$\frac{1}{2} \times 8x = 4x$ and $\frac{1}{2} \times (-6) = -3$. Result: $4x - 3$.

GCF of 36 and 24 is 12. Divide: $36a \div 12 = 3a$ and $24b \div 12 = 2b$. Result: $12(3a + 2b)$.

Subtract 3: $\frac{5x}{2} = 15$. Multiply by 2: $5x = 30$. Divide by 5: $x = 6$. Check: $\frac{5(6)}{2} + 3 = 15 + 3 = 18$ ✓.

Subtract 8: $-3x \geq 15$. Divide by $-3$ and flip: $x \leq -5$.

$\frac{78+85+90+s}{4} \geq 85$. Multiply by 4: $253 + s \geq 340$. Subtract 253: $s \geq 87$. Closed circle at 87, shade right.

The actual room stays the same. Changing the drawing scale does not change the real room's area. The area of the real room is always $12 \times 12 = 144 \ m^2$.

Find more at
ViewMath.com/VA-Grade7

A smaller scale means each cm represents more real distance, so fewer cm are needed. The drawing gets smaller. The shape stays the same.

A vertical cut through a cube perpendicular to the base typically produces a rectangle. If the cut goes through the center parallel to a face, it can be a square.

Supplementary angles sum to $180°$. $110+70 = 180$ ✓. The others: $45+45 = 90$, $60+30 = 90$, $100+100 = 200$. Only C works.

Over the $y$-axis: negate $x$, keep $y$. $(-3,7) \rightarrow (3,7)$.

$\frac{4}{5} = \frac{h}{16}$. Cross-multiply: $5h = 64$. Divide: $h = 12.8$ ft.

Square: $8^2 = 64$ cm². Semicircle: $\frac{1}{2} \times \pi r^2 = \frac{1}{2} \times 3.14 \times 16 = 25.12$ cm². Total: $64 + 25.12 = 89.12$ cm².

$SA = 2(3 \times 2) + 2(3 \times 1) + 2(2 \times 1) = 12 + 6 + 4 = 22$ m². Cans: $22 \div 5 = 4.4$. Round up to 5 cans.

$V = 3 \times 1.5 \times 2 = 9$ ft³.

$2\pi r^2$ gives the area of both circular bases (each base has area $\pi r^2$, and there are 2 of them).

$\frac{8}{200} = 0.04 = 4\%$.

Class 1 center $\approx 10$. Class 2 center $\approx 13$. Difference $\approx 3$ push-ups. Class 2 is higher.

City A: $(62 + 64 + 64 + 66 + 66 + 66 + 68 + 68 + 70 + 72) \div 10 = 666 \div 10 = 66.6$. Range $= 72 - 62 = 10$. City B: $(58 + 60 + 64 + 66 + 68 + 70 + 72 + 74 + 74 + 76) \div 10 = 682 \div 10 = 68.2$. Range $= 76 - 58 = 18$. City A has the smaller range.

*The stem is the tens digit and the leaf is the ones digit, so 5 | 3 means 53.*

*The number 1 appeared 22 times, the most of any number. $P(1) = \frac{22}{120} \approx 0.183$.*

*Outcomes with at least one tail: HT, TH, TT — that is 3 out of 4. $P = \frac{3}{4}$.*

*With 4 equal sections, each should appear about 25 times in 100 spins. The results $(26, 24, 28, 22)$ are all close to 25, suggesting the spinner is fair.*

## ✅ Practice Test 3 — Answer Key

1. A   2. B   3. C   4. B   5. B   6. C   7. C   8. D   9. D   10. B

11. C   12. $x > -3$   13. C   14. 5 in   15. A   16. B   17. C   18. A   19. B

20. C   21. C   22. 126 cm³   23. B   24. B   25. B   26. 2 MADs   27. B   28. D

29. C   30. A

### 💡 Time to Learn! 💡

*Review the explanations below, **especially for the questions you missed**.*

*Understanding why each answer is correct builds stronger problem-solving skills.*

***Tip:*** *Circle any questions you got wrong, then read their explanation carefully.*

## 📖 Practice Test 3 — Detailed Explanations

Find more at
ViewMath.com/VA-Grade7

ViewMath.com

$k = \frac{7}{2} = 3.5$. Equation: $y = 3.5x$. When $x = 15$: $y = 3.5 \times 15 = 52.5$.

$75\% = 0.75$. Multiply: $0.75 \times 32 = 24$ students passed.

Proportion: $\frac{n}{75} = \frac{40}{100}$ gives $n = 30$. Equation: $n = 0.40 \times 75 = 30$. Both give 30.

Change: $13{,}200 - 12{,}000 = 1{,}200$. Percent increase: $1{,}200 \div 12{,}000 = 0.10 = 10\%$.

$\frac{|8-8.5|}{8.5} \times 100 = \frac{0.5}{8.5} \times 100 \approx 5.9\%$.

Different signs: subtract $12 - 7 = 5$. The larger absolute value is 12 (negative), so the answer is $-5$.

Difference $= 134 - (-282) = 134 + 282 = 416$ feet.

Different signs: $6.7 - 4.2 = 2.5$. The larger absolute value (6.7) is negative, so the answer is $-2.5$.

Expand: $7a - 21$. Subtract $2a$: $7a - 21 - 2a = 5a - 21$.

GCF of 20 and 30 is 10. Factor: $20n \div 10 = 2n$ and $-30 \div 10 = -3$. Result: $10(2n - 3)$.

Subtract $4.5$: $-1.5x = -7.5$. Divide by $-1.5$: $x = 5$. Check: $-1.5(5) + 4.5 = -7.5 + 4.5 = -3$ ✓.

Add 4: $-7x < 21$. Divide by $-7$ and flip: $x > -3$.

Divide by $-2$ and flip: $x \geq -5$. Closed circle at $-5$, shade right.

$45 \div 9 = 5$ in.

Get Online

Find more at
ViewMath.com/VA-Grade7

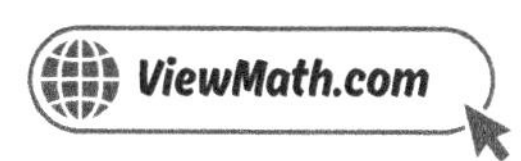

Scale ratio: $\frac{10}{40} = \frac{1}{4}$. Every old length is multiplied by $\frac{1}{4}$ on the new drawing.

A vertical cut perpendicular to the triangular bases of a triangular prism produces a rectangle (the cut goes through the lateral faces).

Adjacent angles are supplementary: $(3y + 10) + (5y - 30) = 180$. Combine: $8y - 20 = 180$. Solve: $8y = 200$, $y = 25$.

From $P(1, 2)$ to $P'(3, 0)$: $x$ changed by $+2$ (right 2) and $y$ changed by $-2$ (down 2).

$\frac{5}{3} = \frac{h}{15}$. Cross-multiply: $3h = 75$. Divide: $h = 25$ ft.

Horizontal: $12 \times 2 = 24$ cm$^2$. Vertical: $2 \times 8 = 16$ cm$^2$. Total: $24 + 16 = 40$ cm$^2$.

Surface area depends on the square of the dimensions. Doubling all dimensions multiplies each face area by $2^2 = 4$, so the total surface area quadruples.

Bottom prism: $10 \times 3 \times 3 = 90$ cm$^3$. Top prism: $4 \times 3 \times 3 = 36$ cm$^3$. Total: $90 + 36 = 126$ cm$^3$.

$V = \pi r^2 h$. If $h$ becomes $2h$, then $V = \pi r^2 (2h) = 2\pi r^2 h$. The volume doubles.

Larger random samples better represent the population and give more reliable results.

Difference in medians: $88 - 82 = 6$. In MADs: $\frac{6}{4} = 1.5$ MADs.

Difference: $57 - 45 = 12$. In MADs: $\frac{12}{6} = 2$.

The key tells readers what the stems and leaves represent, so they can correctly read the data values.

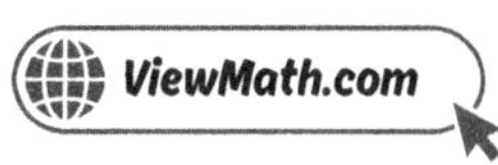

$P = \frac{18}{90} = \frac{1}{5} = 0.2$. Both A and C represent the same value.

The only way to get a sum of 2 is $(1, 1)$. $P = \frac{1}{36}$.

$P = \frac{6}{40} = 0.15$. The theoretical probability is $\frac{1}{8} = 0.125$, close to the simulation result.

---

## ✅ Practice Test 4 — Answer Key

| | | | | | | | | |
|---|---|---|---|---|---|---|---|---|
| **1** B | **2** 80% | **3** D | **4** 15% | **5** C | **6** D | **7** C | **8** $-\frac{1}{10}$ | **9** A |
| **10** B | **11** B | **12** $\frac{x}{4} - 3 \geq 2;\ x \geq 20$ | **13** A | **14** $1\ cm = 30\ m$ | **15** C | **16** B | **17** C |
| **18** A | **19** A | **20** B | **21** C | **22** B | **23** C |

**24** Random variation (or the sample may be biased)   **25** C   **26** B   **27** C   **28** C   **29** $\frac{3}{8}$

**30** B

---

### 💡 Time to Learn! 💡

Review the explanations below, **especially for the questions you missed.**

Understanding why each answer is correct builds stronger problem-solving skills.

*Tip:* Circle any questions you got wrong, then read their explanation carefully.

---

## 📖 Practice Test 4 — Detailed Explanations

The cost is proportional to hours at \$8/hr, so $c = 8h$.

Divide: $280 \div 350 = 0.80 = 80\%$.

$\frac{135}{w} = \frac{54}{100}$. Cross-multiply: $54w = 13{,}500$. Divide: $w = 250$.

Change: $400 - 340 = 60$. Percent decrease: $60 \div 400 = 0.15 = 15\%$.

$\frac{|50-40|}{40} \times 100 = \frac{10}{40} \times 100 = 25\%$.

A number plus its opposite (additive inverse) always equals $0$.

$4 - (-11) = 4 + 11 = 15°F$.

Common denominator 10: $\frac{7}{10} - \frac{8}{10} = -\frac{1}{10}$.

Expand each: $3x - 6 + 2x + 8$. Combine: $(3x + 2x) + (-6 + 8) = 5x + 2$.

GCF of $6x$ and 12 is 6. Factor: $6x \div 6 = x$ and $12 \div 6 = 2$. Result: $6(x + 2)$.

Multiply every term by 6: $3n + 2n = 60$. Combine: $5n = 60$. Divide by 5: $n = 12$. Check: $\frac{12}{2} + \frac{12}{3} = 6 + 4 = 10$ ✓.

$\frac{x}{4} - 3 \geq 2$. Add 3: $\frac{x}{4} \geq 5$. Multiply by 4: $x \geq 20$.

Solve: $2x > 8$, so $x > 4$. This is an open circle at 4, shade right — Graph A.

$360 \div 12 = 30$. The scale is $1\ cm = 30\ m$.

Find more at
ViewMath.com/VA-Grade7

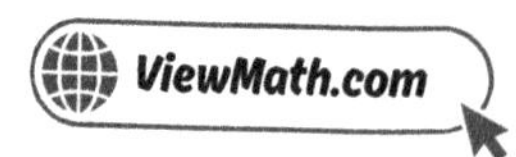

Scale ratio: $\frac{20}{10} = 2$. Every length on the new drawing is 2 times the old one. The drawing gets bigger because each cm now represents fewer real km.

A cut parallel to the base of a prism gives a shape congruent to the base. Since the base is a triangle, the cross-section is a triangle.

$3x - 5 + x + 15 = 90$. Combine: $4x + 10 = 90$. Solve: $4x = 80$, $x = 20$. Angle A: $3(20) - 5 = 55°$.

Reflection over the $x$-axis negates $y$. To reverse: $K'(4, -3) \rightarrow K(4, 3)$.

Scale factor: $\frac{5}{10} = 0.5$. $YZ = 14 \times 0.5 = 7$.

Rectangle: $18 \times 6 = 108$ in². Triangle: $\frac{1}{2} \times 6 \times 4 = 12$ in². Banner: $108 - 12 = 96$ in².

A rectangular prism has 6 faces: top, bottom, front, back, left, and right.

Volume measures three-dimensional space and is measured in cubic units.

Two circles: $2 \times \pi r^2 = 2 \times 3.14 \times 25 = 157$ cm². Rectangle: $31.4 \times 12 = 376.8$ cm². Total: $157 + 376.8 = 533.8$ cm².

Even a well chosen random sample may differ slightly from the population due to chance.

A complete comparison requires looking at both center (where the data clusters) and spread (how spread out it is).

Difference $= \$620 - \$500 = \$120$. In MADs: $\frac{120}{40} = 3$ MADs.

With an even number of values (12), the median is the average of the 6th and 7th values.

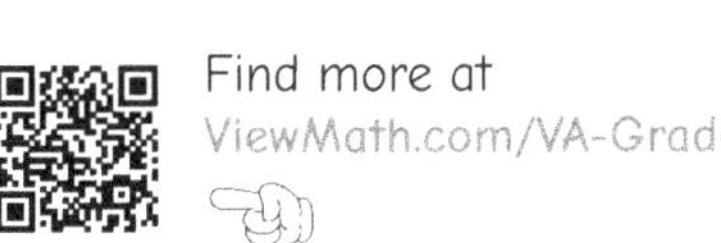
Find more at
ViewMath.com/VA-Grade7

The Law of Large Numbers states that as trials increase, experimental probability approaches theoretical probability.

Outcomes with exactly 2 tails: HTT, THT, TTH — that is 3 out of 8. $P = \frac{3}{8}$.

Each coin flip has a 50% chance of heads (pass). Flipping 3 coins and counting exactly 2 heads simulates exactly 2 passes.

## ✅ Practice Test 5 — Answer Key

**1** $y = 3.5x;\ x = 14$  **2** C  **3** D  **4** A  **5** D  **6** $-5$  **7** B  **8** B  **9** C

**10** $3x + 2$  **11** $x = 12$  **12** B  **13** $x > 2$; open circle at 2, shade right  **14** B  **15** C

**16** A triangle  **17** C  **18** B  **19** B  **20** A  **21** 8 in  **22** C  **23** C  **24** B

**25** A  **26** B  **27** 81  **28** B  **29** $\frac{1}{9}$  **30** C

## 💡 Time to Learn! 💡

*Review the explanations below, **especially for the questions you missed**.*

*Understanding why each answer is correct builds stronger problem-solving skills.*

***Tip:*** *Circle any questions you got wrong, then read their explanation carefully.*

## 📖 Practice Test 5 — Detailed Explanations

**1** $k = \frac{17.5}{5} = 3.5$. Equation: $y = 3.5x$. When $y = 49$: $49 = 3.5x$, so $x = 14$.

Find more at
ViewMath.com/VA-Grade7

Divide the part by the whole: $18 \div 72 = 0.25 = 25\%$.

Boys: $0.45 \times 360 = 162$. Girls: $360 - 162 = 198$. Alternatively, girls are $55\%$: $0.55 \times 360 = 198$.

April to May: $(500 - 400) \div 400 = 0.25 = 25\%$. May to June: $(550 - 500) \div 500 = 0.10 = 10\%$. The greatest percent change is $25\%$.

$\frac{|2.4 - 2|}{2} \times 100 = \frac{0.4}{2} \times 100 = 20\%$.

Group positives: $5 + 3 = 8$. Group negatives: $(-11) + (-2) = -13$. Combine: $8 + (-13) = -5$.

Subtraction means adding the opposite: $(-5) - 3 = (-5) + (-3)$.

$-\frac{3}{8} + n = \frac{1}{2} = \frac{4}{8}$. So $n = \frac{4}{8} - \left(-\frac{3}{8}\right) = \frac{4}{8} + \frac{3}{8} = \frac{7}{8}$.

$-5 \times x = -5x$ and $-5 \times 2 = -10$. Result: $-5x - 10$.

Area = width $\times$ length. So length $= \frac{18x + 12}{6}$. Factor: $18x + 12 = 6(3x + 2)$. Length $= 3x + 2$.

Add 2: $\frac{3x}{4} = 9$. Multiply by 4: $3x = 36$. Divide by 3: $x = 12$. Check: $\frac{3(12)}{4} - 2 = 9 - 2 = 7$ ✓.

Add 10: $5x \geq 35$. Divide by 5: $x \geq 7$.

Subtract 10: $-4x < -8$. Divide by $-4$ and flip: $x > 2$. Open circle at 2, shade right.

$3.2 \times 50 = 160$ km.

Scale ratio: $\frac{5}{2} = 2.5$. New dimensions: $3 \times 2.5 = 7.5$ cm and $4 \times 2.5 = 10$ cm. New area: $7.5 \times 10 = 75$ cm$^2$.

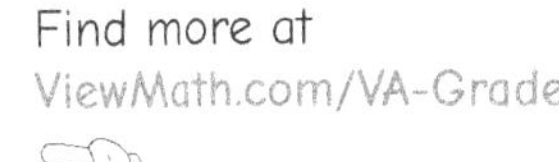
Find more at
ViewMath.com/VA-Grade7

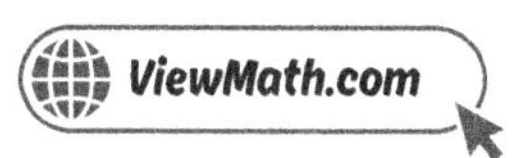

A vertical slice through the apex of a cone produces an isosceles triangle. The base of the triangle is a diameter of the cone's base.

If two equal angles sum to $180°$, then each is $180 \div 2 = 90°$.

Reflection over the $y$-axis: negate $x$, keep $y$. $(5, -3) \rightarrow (-5, -3)$.

Scale factor: $\frac{20}{5} = 4$. Shortest side: $3 \times 4 = 12$.

Room: $14 \times 10 = 140 \ ft^2$. Closet: $3 \times 4 = 12 \ ft^2$. Remaining: $140 - 12 = 128 \ ft^2$.

$s^2 = 384 \div 6 = 64$. So $s = 8 \ in$.

$V = 50 \times 30 \times 40 = 60.000 \ cm^3$.

$V = \pi r^2 h = 3.14 \times 36 \times 14 = 1.582.56 \ m^3$.

The population is the entire group you want to study — all 800 students.

A smaller MAD means scores are closer to the mean, so Team A is more consistent.

Same mean means same center. Different MADs mean one group's data is more spread out than the other's.

Data: $70, 73, 75, 78, 81, 84, 86, 90, 95$. There are 9 values; the 5th is 81.

For a fair spinner, each section would appear about 30 times. The results (35, 28, 27) are close enough to suggest the spinner is likely fair.

Products of 6: $(1,6),(2,3),(3,2),(6,1)$ — 4 outcomes. $P = \frac{4}{36} = \frac{1}{9}$.

Before choosing a model or running trials, you must first identify what event you're simulating and what the possible outcomes are.

## ✅ Practice Test 6 — Answer Key

$c = 7.50t$    D    C    B    C    C    C    B    B

B    A    B    $x = 3$    A    C    C    137°    $(1,5)$

B    50 $m^2$    C    240 $m^3$    C    C    B    B    B

A    C    B

### 💡 Time to Learn! 💡

Review the explanations below, **especially for the questions you missed.**

Understanding why each answer is correct builds stronger problem-solving skills.

**Tip:** Circle any questions you got wrong, then read their explanation carefully.

## 📖 Practice Test 6 — Detailed Explanations

$k = \frac{22.50}{3} = 7.50$ dollars per ticket. The equation is $c = 7.50t$.

$40\% = 0.40$. Multiply: $0.40 \times 150 = 60$.

Cross-multiply: $100 \times n = 250 \times 16$, which gives $100n = 4000$.

Multiply by the decrease multiplier: $200 \times 0.65 = \$130$.

A: $\frac{|180-200|}{200} = 10\%$. B: $\frac{|220-200|}{200} = 10\%$. Both have 10% error.

Different signs: $14 - 9 = 5$. The larger absolute value (14) is negative, so the answer is $-5$.

Distance $= |-9 - 4| = |-13| = 13$ units. Distance is always positive.

$\frac{3}{4} - \frac{3}{2} = \frac{3}{4} - \frac{6}{4} = -\frac{3}{4}$.

$-2 \times 3x = -6x$ and $-2 \times 7 = -14$. Result: $-6x - 14$.

GCF is 3. Factor out $-3$: $-9x \div (-3) = 3x$ and $-6 \div (-3) = 2$. Result: $-3(3x + 2)$. Check: $-3(3x + 2) = -9x - 6$ ✓.

Subtract 5.6: $-0.2n = -1.6$. Divide by $-0.2$: $n = 8$. Check: $-0.2(8) + 5.6 = -1.6 + 5.6 = 4$ ✓.

An open circle at 5 with shading to the right means $x > 5$ (not including 5).

Add 3: $7x \geq 21$. Divide by 7: $x \geq 3$. The smallest integer solution is 3.

Divide the actual length by the scale factor: $180 \div 24 = 7.5$ inches.

Scale ratio: $\frac{12}{4} = 3$. New width: $0.5 \times 3 = 1.5$ in.

Get Online

Find more at
ViewMath.com/VA-Grade7

ViewMath.com

Any cut parallel to the base of a prism gives a shape congruent to the base. A hexagonal prism has a hexagonal base, so the cross-section is a hexagon.

$180° - 43° = 137°.$

Reflect over $x$-axis: $(-3, -5) \to (-3, 5)$. Translate right 4: $(-3 + 4, 5) = (1, 5)$.

Smaller side $= 30 \times \frac{2}{5} = 12 \ cm.$

$A = \frac{1}{2}(b_1 + b_2) \times h = \frac{1}{2}(8 + 12) \times 5 = \frac{1}{2} \times 20 \times 5 = 50 \ m^2.$

$SA = 2(6 \times 6) + 2(6 \times 10) + 2(6 \times 10) = 72 + 120 + 120 = 312 \ ft^2.$

$B = \frac{1}{2} \times 10 \times 6 = 30 \ m^2. \ V = 30 \times 8 = 240 \ m^3.$

$628 = 3.14 \times r^2 \times 8 = 25.12 \, r^2. \ r^2 = 628 \div 25.12 = 25. \ r = 5 \ cm.$

In a random sample, every member of the population has an equal chance of being chosen.

Dot plots show every data point, which is useful for small data sets where individual values matter.

There are 7 values. The middle value (4th) is 20.

A back-to-back plot has leaves for one group on the left and for another group on the right, sharing the same stems.

$P = \frac{red\ draws}{total\ draws} = \frac{14}{40} = 0.35.$

Find more at
ViewMath.com/VA-Grade7

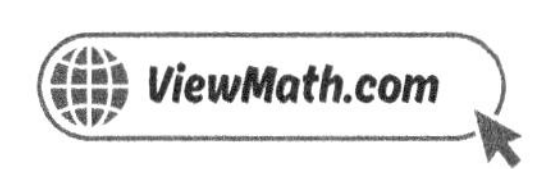

$P(\text{tails}) = \frac{1}{2}$ and $P(\text{even}) = \frac{3}{6} = \frac{1}{2}$. $P = \frac{1}{2} \times \frac{1}{2} = \frac{1}{4}$. Or: 3 favorable out of 12 total outcomes.

A simulation uses random numbers, coins, dice, or spinners to model a real-world event and estimate probabilities.

## ☑ Practice Test 7 — Answer Key

**1.** $y = 4.5x$; when $x = 20$, $y = 90$  **2.** $C$  **3.** 15%  **4.** $B$  **5.** $\approx 11.1\%$  **6.** $B$

**7.** 21,500 *feet*  **8.** $B$  **9.** $B$  **10.** $B$  **11.** $p = \$60$  **12.** $A$  **13.** $A$  **14.** $C$

**15.** 15 *cm*  **16.** $B$  **17.** $C$  **18.** $A$  **19.** $B$  **20.** 52 $cm^2$  **21.** $C$  **22.** $C$

**23.** 1,130.4 $cm^3$  **24.** $B$  **25.** Class A  **26.** $B$

**27.** Stem 7: 5 8 9; Stem 8: 1 3 5 7 8; Stem 9: 0 2. Mean = 83.8.  **28.** $B$  **29.** $B$  **30.** $C$

## 💡 Time to Learn! 💡

Review the explanations below, **especially for the questions you missed**.

Understanding why each answer is correct builds stronger problem-solving skills.

**Tip:** Circle any questions you got wrong, then read their explanation carefully.

## 📖 Practice Test 7 — Detailed Explanations

**1.** $k = \frac{18}{4} = 4.5$. Equation: $y = 4.5x$. At $x = 20$: $y = 4.5 \times 20 = 90$.

Find more at
ViewMath.com/VA-Grade7

*Divide the part by the percent: $36 \div 0.45 = 80$.*

*$\frac{48}{320} = \frac{p}{100}$. Cross-multiply: $320p = 4800$. Divide: $p = 15\%$.*

*Change: $75 - 60 = 15$. Percent decrease: $15 \div 75 = 0.20 = 20\%$.*

*$\frac{|80-72|}{72} \times 100 = \frac{8}{72} \times 100 \approx 11.1\%$.*

*Same signs: add the absolute values $8 + 5 = 13$ and keep the negative sign: $-13$.*

*$30{,}000 - 8{,}500 = 21{,}500$ feet.*

*$-\frac{7}{8} + \left(-\frac{1}{4}\right) = -\frac{7}{8} + \left(-\frac{2}{8}\right) = -\frac{9}{8}$.*

*Multiply 3 by each term: $3 \times x = 3x$ and $3 \times 4 = 12$. Result: $3x + 12$.*

*GCF of 16, 24, and 8 is 8. Divide each: $16x \div 8 = 2x$, $24y \div 8 = 3y$, $-8 \div 8 = -1$. Result: $8(2x + 3y - 1)$.*

*After $\frac{1}{4}$ markdown: $p - \frac{p}{4} = \frac{3p}{4}$. After subtracting \$8: $\frac{3p}{4} - 8 = 37$. Add 8: $\frac{3p}{4} = 45$. Multiply by $\frac{4}{3}$: $p = 60$.*

*Current height plus growth: $42 + g > 48$.*

*Subtract 3: $4x \le 16$. Divide by 4: $x \le 4$. Closed circle at 4, shade left.*

*A scale drawing keeps all lengths proportional using the same scale factor. Angles are preserved, and area scales by $k^2$, not $k$.*

*Scale ratio: $\frac{12}{4} = 3$. New length: $5 \times 3 = 15$ cm.*

Find more at
ViewMath.com/VA-Grade7

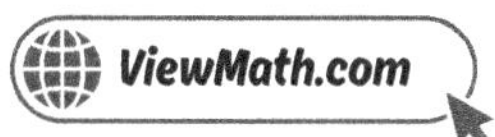
ViewMath.com

*A horizontal cut through a cylinder parallel to the circular base produces a circle.*

*Supplementary angles sum to $180°$. $180° - 115° = 65°$.*

*Rotating the origin about itself keeps it at $(0, 0)$. Any rotation of a point about itself results in no change.*

*Scale factor: $\frac{15}{6} = 2.5$. Shorter side: $4 \times 2.5 = 10$ in.*

*Top: $10 \times 2 = 20$ cm$^2$. Bottom: $10 \times 2 = 20$ cm$^2$. Middle: $2 \times 6 = 12$ cm$^2$. Total: $20 + 20 + 12 = 52$ cm$^2$.*

*$SA = 2(30 \times 20) + 2(30 \times 6) + 2(20 \times 6) = 1200 + 360 + 240 = 1,800$ cm$^2$.*

*$V = 2 \times 1.5 \times 3 = 9$ ft$^3$.*

*$V = \pi r^2 h = 3.14 \times 36 \times 10 = 1,130.4$ cm$^3$.*

*Selecting every 10th student from the school roster is systematic and gives all students a chance, making it the least biased.*

*Class A has a smaller IQR (6 vs. 14), meaning the middle 50% of scores are more tightly clustered.*

*A wider range indicates more variability, not better performance. Class A has a higher median and is more consistent (smaller IQR and range).*

*Ordered: $75, 78, 79, 81, 83, 85, 87, 88, 90, 92$. Sum $= 838$. Mean $= \frac{838}{10} = 83.8$.*

*$P = \frac{112}{200} = 0.56$.*

$P(heads) = \frac{1}{2}$. Numbers greater than 4: 5, 6, so $P = \frac{2}{6} = \frac{1}{3}$. $P = \frac{1}{2} \times \frac{1}{3} = \frac{1}{6}$.

A spinner with 4 equal sections has $\frac{1}{4} = 25\%$ chance for each section. Assigning one section as "win" gives 25%.

## ✔ Practice Test 8 — Answer Key

1. C
2. 90 *students*
3. 140
4. D
5. $\approx 4.2\%$
6. B
7. C
8. D
9. B
10. $4(5x + 3) = 4 \times 5x + 4 \times 3 = 20x + 12$
11. A
12. B
13. $x \geq -6$
14. 280 *m*
15. 1 *cm* = 18 *ft*
16. Smaller than the base
17. $x = 6$; the angles are 43°
18. B
19. 20
20. B
21. 432 *in*$^2$
22. B
23. C
24. 360
25. Yes; the means are about 2.5 to 3.3 MADs apart
26. B
27. 11
28. B
29. $\frac{1}{12}$
30. B

## 💡 Time to Learn! 💡

*Review the explanations below, especially for the questions you missed.*

*Understanding why each answer is correct builds stronger problem-solving skills.*

*Tip: Circle any questions you got wrong, then read their explanation carefully.*

## 📖 Practice Test 8 — Detailed Explanations

$y = 6 \times 7 = 42$.

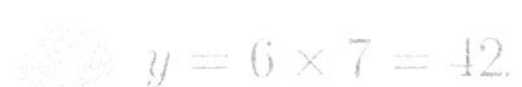

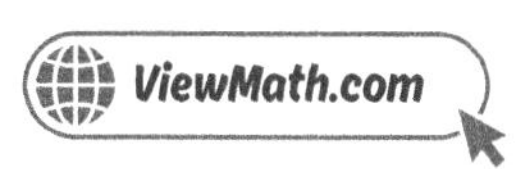

*45% of 200:* $0.45 \times 200 = 90$ *students take the bus.*

$\frac{91}{w} = \frac{65}{100}$. *Cross-multiply:* $65w = 9100$. *Divide:* $w = 140$.

*Change:* $54 - 45 = 9$. *Percent increase:* $9 \div 45 = 0.20 = 20\%$.

$\frac{|250-240|}{240} \times 100 = \frac{10}{240} \times 100 \approx 4.17 \approx 4.2\%$.

$0 + 30 + (-45) + 20 + (-15) = 30 + 20 - 45 - 15 = 50 - 60 = -10$. *The balance is* $-\$10$.

$(-1) - (-9) = (-1) + 9 = 8$, *which is positive.*

$-6.4 + 9.2 = 2.8°C$. *Different signs:* $9.2 - 6.4 = 2.8$, *positive.*

*Expand:* $2x + 6$. *Add 5:* $2x + 6 + 5 = 2x + 11$.

*Distribute 4:* $4 \times 5x = 20x$ *and* $4 \times 3 = 12$. *So* $4(5x + 3) = 20x + 12$ ✓. *They are equivalent.*

*If* $\frac{2}{3}x = 8$, *then each* $\frac{1}{3}$ *of* $x$ *is 4. So* $x = 3 \times 4 = 12$.

*Subtract 7:* $3x \leq 18$. *Divide by 3:* $x \leq 6$.

*Closed circle means* $-6$ *is included, shading right means values greater than or equal to* $-6$: $x \geq -6$.

*Each square is 10 m. Length:* $9 \times 10 = 90$ *m. Width:* $5 \times 10 = 50$ *m. Perimeter:* $2(90 + 50) = 2(140) = 280$ *m.*

*Actual length:* $9 \times 6 = 54$ *ft. New scale:* $54 \div 3 = 18$ *ft per cm.*

A horizontal cut parallel to the base of a pyramid produces a similar shape that is smaller than the base. Halfway up, the cross-section is a 4 cm square (half the edge length).

Vertical angles are equal: $7x + 1 = 5x + 13$. Subtract $5x$: $2x + 1 = 13$. Subtract 1: $2x = 12$. Divide: $x = 6$. Angle: $7(6) + 1 = 43°$.

Translations, reflections, and rotations are all rigid motions — they preserve size and shape. Reflections reverse orientation, and rotations change direction, but all keep the figure congruent.

Scale factor: $\frac{12.5}{5} = 2.5$. $XY = 8 \times 2.5 = 20$.

Rectangle: $10 \times 4 = 40\ m^2$. Triangle: $\frac{1}{2} \times 10 \times 3 = 15\ m^2$. Total: $40 + 15 = 55\ m^2$.

$SA = 2(12 \times 8) + 2(12 \times 6) + 2(8 \times 6) = 192 + 144 + 96 = 432\ in^2$.

The general formula for the volume of any prism is $V = Bh$, where $B$ is the area of the base and $h$ is the height.

$r = 4\ cm$. $V = \pi r^2 h = 3.14 \times 16 \times 5 = 251.2\ cm^3$.

$\frac{6}{50} = 12\%$. Then 12% of $3,000 = 360$.

Difference $= 10$. Using MAD of 4: $\frac{10}{4} = 2.5$. Using MAD of 3: $\frac{10}{3} \approx 3.3$. Both are greater than 2, so the difference is meaningful.

$Q_1 = 22$, $Q_3 = 32$. $IQR = 32 - 22 = 10$.

Count leaves: $3 + 5 + 2 + 1 = 11$.

Find more at
ViewMath.com/VA-Grade7

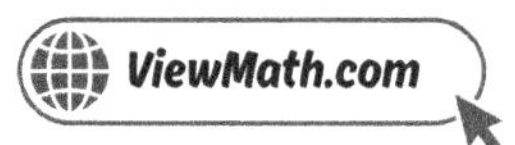

$P = \frac{16}{52} = \frac{4}{13} \approx 0.308$. Note: $\frac{16}{52}$ and $\frac{4}{13}$ are equivalent, but B directly shows the experimental data.

$P(tails) = \frac{1}{2}$, $P(1) = \frac{1}{6}$. $P = \frac{1}{2} \times \frac{1}{6} = \frac{1}{12}$.

More trials produce more reliable estimates. The experimental probability approaches the theoretical value as trials increase.

## ✅ Practice Test 9 — Answer Key

**1** B    **2** B    **3** D    **4** B    **5** A    **6** A    **7** A    **8** $-0.4°C$    **9** B

**10** $h = 4$    **11** B    **12** A    **13** B    **14** B    **15** A    **16** B    **17** B    **18** B    **19** C

**20** B    **21** C    **22** D    **23** B    **24** C    **25** D    **26** B    **27** B    **28** A    **29** C

**30** Roll the die; let 1 and 2 represent the event (or any 2 numbers out of 6)

## 💡 Time to Learn! 💡

Review the explanations below, **especially for the questions you missed**.

Understanding why each answer is correct builds stronger problem-solving skills.

**Tip:** Circle any questions you got wrong, then read their explanation carefully.

## 📖 Practice Test 9 — Detailed Explanations

$k = \frac{52}{8} = 6.5$, so $y = 6.5x$.

Convert to a decimal: $25\% = 0.25$. Multiply: $0.25 \times 80 = 20$.

$\frac{72}{w} = \frac{90}{100}$. Cross-multiply: $90w = 7200$. Divide: $w = 80$.

A 40% decrease means you keep $100\% - 40\% = 60\%$. The multiplier is $0.60$.

If estimated = actual, then $|estimated - actual| = 0$, so percent error = 0%.

Add the negatives first: $(-3) + (-9) = -12$. Then $(-12) + 4 = -8$.

$-25 + 10 = -15$. Then $-15 - 7 = -15 + (-7) = -22$ feet.

Start at $-2.5$. Rise 3.8: $-2.5 + 3.8 = 1.3°C$. Drop 1.7: $1.3 - 1.7 = -0.4°C$.

$-3 \times (-4)$ should equal $+12$, not $-12$. The correct expansion is $-3x + 12$.

Total area $= h(2x) + h(3) = h(2x + 3) = 8x + 12$. Factor: $8x + 12 = 4(2x + 3)$. So $h = 4$.

Multiply every term by 6. $x - 4 = 3$. Add 4: $x = 7$. Check: $\frac{7}{6} - \frac{2}{3} = \frac{7}{6} - \frac{4}{6} = \frac{3}{6} = \frac{1}{2}$ ✓.

Subtract 3: $2x > 8$. Divide by 2: $x > 4$.

$x < 5$ uses an open circle (not including 5) while $x \leq 5$ uses a closed circle (including 5). Both shade left.

$3.5 \times 6 = 21$ ft.

Scale ratio: $\frac{25}{75} = \frac{1}{3}$. New length: $6 \times \frac{1}{3} = 2$ cm.

Find more at
ViewMath.com/VA-Grade7

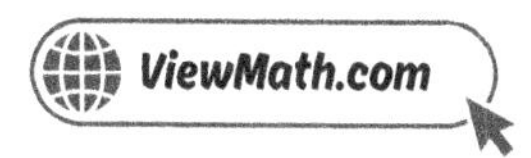

A diagonal plane that passes through all six faces of a cube creates a regular hexagonal cross-section.

Vertical angles are equal: $5x - 20 = 3x + 10$. Subtract $3x$: $2x - 20 = 10$. Add 20: $2x = 30$. Divide $x = 15$. Angle: $5(15) - 20 = 55°$.

90° clockwise: $(x, y) \rightarrow (y, -x)$. $(2, 6) \rightarrow (6, -2)$.

Scale factor: $\frac{21}{7} = 3$. Area ratio: $3^2 = 9$. So the ratio is $1 : 9$.

Rectangle: $8 \times 6 = 48$ $in^2$. Two semicircles = one full circle with $r = 3$: $3.14 \times 9 = 28.26$ $in^2$. Total: $48 + 28.26 = 76.26$ $in^2$.

$94 = 2(5 \times 3) + 2(5 \times h) + 2(3 \times h) = 30 + 10h + 6h = 30 + 16h$. So $16h = 64$, $h = 4$ cm.

$V = lwh = 6 \times 4 \times 3 = 72$ $cm^3$.

$r = 5$ in. $V = \pi r^2 h = 3.14 \times 25 \times 6 = 471$ $in^3$.

The sample is the smaller group chosen from the population — the 100 students measured.

Class A median = 78, Class B median = 65. Class A's range (60–95) overlaps with Class B's range (50–80), but Class A is higher overall.

Group A range = 20, Group B range = 60. Group B's data is far more spread out.

Class A data (9 values): $62, 65, 70, 73, 75, 78, 84, 86, 92$. Median = 75. Class B data (12 values): $63, 65, 68, 72, 74, 79, 81, 83, 85, 87, 90, 95$. Median $= \frac{79+81}{2} = 80$. Class B has a higher median.

Experimental: $\frac{25}{80} = 0.3125$. Theoretical: $\frac{1}{4} = 0.25$. The experimental is slightly higher.

Find more at
ViewMath.com/VA-Grade7

Only one outcome out of 8 is all tails (TTT). $P = \frac{1}{8}$.

Choose 2 of the 6 faces to represent the event. $P = \frac{2}{6} = \frac{1}{3}$.

## Practice Test 10 — Answer Key

B    C    C    C    B    $-2°C$    7    8.4 meters    C

B    $6    B    $x < 3$; open circle at 3, shade left    C    C

A circle (the largest possible circle)    A    $A'(1,-1)$, $B'(5,-1)$, $C'(5,-3)$, $D'(1,-3)$    C

A    C    C    $157\ m^3$    C    B    Median $= 300$; IQR $= 200$

B    A    B    A

## Time to Learn!

Review the explanations below, **especially for the questions you missed.**

Understanding why each answer is correct builds stronger problem-solving skills.

**Tip:** Circle any questions you got wrong, then read their explanation carefully.

## Practice Test 10 — Detailed Explanations

From the point $(4, 5)$: $k = \frac{5}{4} = 1.25$. The equation is $c = 1.25p$.

Find more at
ViewMath.com/VA-Grade7

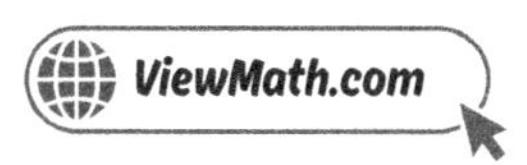

The grid has 100 squares total. Count the shaded squares: 3 full rows of $10 = 30$, plus 5 more $= 35$. So 35% is shaded.

Set up $\frac{n}{120} = \frac{35}{100}$. Cross-multiply: $100n = 4200$. Divide: $n = 42$.

A: $100 \times 1.30 = 130$. B: $100 \times 1.20 \times 1.10 = 132$. C: $100 \times 1.15 \times 1.15 = 132.25$. C is the greatest.

First: $\frac{2}{12} \times 100 \approx 16.7\%$. Second: $\frac{2}{52} \times 100 \approx 3.8\%$. The second has a smaller percent error.

Start at $-3°C$. Rise $4°$: $-3 + 4 = 1°C$. Drop $3°$: $1 + (-3) = -2°C$.

$(-8) - (-15) = (-8) + 15 = 7$.

Change: $-23.8 - (-15.4) = -23.8 + 15.4 = -8.4$. The hiker descended 8.4 meters.

$-4 \times 2y = -8y$ and $-4 \times (-5) = +20$. A negative times a negative is positive. Result: $-8y + 20$.

GCF of 8 and 20 is 4. Factor: $8m \div 4 = 2m$ and $20 \div 4 = 5$. Result: $4(2m + 5)$. Choice A uses 2, which is not the GCF.

Student C $= \frac{1}{2} \times 12 = \$6$. Check: Student B $= 31.50 - 12 - 6 = \$13.50$. Total: $12 + 13.50 + 6 = 31.50$ ✓.

Subtract 7: $-2x < -6$. Divide by $-2$ and flip: $x > 3$.

Subtract 9: $-3x > -9$. Divide by $-3$ and flip: $x < 3$. Open circle at 3, shade left.

Actual dimensions: $6 \times 8 = 48$ m and $2.5 \times 8 = 20$ m. Perimeter: $2(48 + 20) = 2(68) = 136$ m.

Actual road: $7 \times 5 = 35$ km. New scale: $35 \div 3.5 = 10$. So the second map uses 1 cm = 10 km.

A cut through the center of a sphere produces a great circle — the largest possible circular cross-section, with the same radius as the sphere.

Vertical angles are equal ($130°$ and $130°$). Adjacent angles are supplementary: $180° - 130° = 50°$. The four angles are $130°$, $50°$, $130°$, $50°$.

Over the $x$-axis: keep $x$, negate $y$. $(1,1) \to (1,-1)$, $(5,1) \to (5,-1)$, $(5,3) \to (5,-3)$, $(1,3) \to (1,-3)$.

$\frac{60}{10} = \frac{h}{12}$. Cross-multiply: $40h = 720$. Divide: $h = 18$ ft.

Yard: $20 \times 15 = 300$ m$^2$. Circle: $3.14 \times 9 = 28.26$ m$^2$. Remaining: $300 - 28.26 = 271.74$ m$^2$.

A cube has 6 faces. $SA = 6s^2 = 6 \times 16 = 96$ cm$^2$.

$B = \frac{1}{2} \times 5 \times 12 = 30$ cm$^2$. $V = 30 \times 15 = 450$ cm$^3$.

$V = \pi r^2 h = 3.14 \times 25 \times 2 = 157$ m$^3$.

Average the two samples: $\frac{18+22}{2} = 20$ out of $50 = 40\%$. Then $40\%$ of $1,000 = 400$.

The median is less affected by skew than the mean, so it is the best measure to compare a skewed and a symmetric distribution.

Median (3rd value) $= 300$. $Q_1 = 200$, $Q_3 = 400$. $IQR = 400 - 200 = 200$.

Leaves must be arranged in order from least to greatest so the data is easy to read.

Find more at
ViewMath.com/VA-Grade7

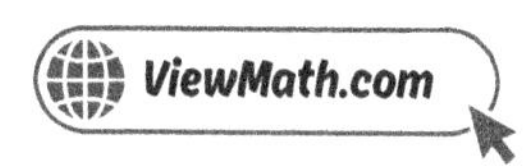

 The bar for $3$ reaches $6$. $P = \frac{6}{60} = \frac{1}{10}$.

Pairs with product $12$: $(2,6), (6,2), (3,4), (4,3)$ — that is $4$ out of $36$. $P = \frac{4}{36} = \frac{1}{9}$.

$P(\text{sum} = 7) = \frac{6}{36} = \frac{1}{6}$. Expected: $36 \times \frac{1}{6} = 6$. The simulation gave $8$, slightly above expected.

Well done checking your answers!

# Author's Final Note

I hope you enjoyed this book as much as I enjoyed writing it. Whether you are a student working through the material, a parent supporting your child's learning, or a teacher guiding your class, I have tried to make this book as clear and engaging as possible. I hope I have succeeded. If you have any suggestions for improvement, please let me know. I would love to hear from you.

The accuracy of calculations is very important to me. We have done our best, but I also expect that I have made some minor errors. Constant improvement is the name of the game. If you find any errors, please let me know. I will fix them in the next edition.

**For students:** Your learning journey does not end here. I have written a series of books to help you learn math. Make sure you browse through them. I especially recommend workbooks and practice tests to help you prepare for your exams.

**For parents:** Thank you for investing in your child's education. I encourage you to explore the companion resources available online to help support your child outside the classroom.

**For teachers:** Thank you for the invaluable work you do every day. I hope this book serves as a useful resource in your classroom. Feel free to reach out if you have suggestions or would like to discuss how best to use this book with your students.

I also enjoy reading your reviews. If you have a moment, please leave a review on where you found this book. It will help others find this book. If you have any questions or comments, please feel free to contact me at drNazari@ViewMath.com.

And one last thing: Remember to use online resources for additional help. I recommend using the resources on https://ViewMath.com You can find video lessons, practice problems, and more. You can also use the online companion for this book to track your progress and access additional resources.

Wishing all students the best in their studies, parents every success in supporting their children, and teachers continued inspiration in their classrooms!

Dr. A. Nazari

Find more at
ViewMath.com/VA-Grade7

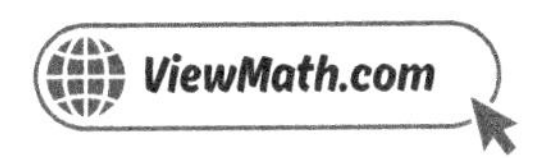

 # Great Job! Keep Learning with ViewMath!

Keep up the great work! Visit **viewmath.com/VA-Grade7** for free lessons, quizzes, and more.

Study Guide

Workbook

Step-by-Step

3 Practice Tests

5 Practice Tests

7 Practice Tests

Find more at
ViewMath.com/VA-Grade7

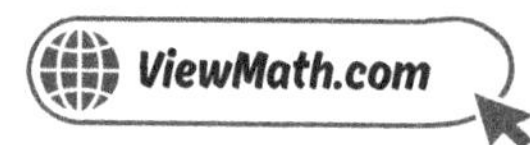

ViewMath.com

www.ingramcontent.com/pod-product-compliance
Lightning Source LLC
Chambersburg PA
CBHW081146130726
47996CB00009B/3009